A RECKLESS GOD?

Currents and Challenges in the Christian Conversation with Science

Edited by
Roland Ashby, Chris Mulherin,
John Pilbrow and Stephen Ames

a. Acorn Press

Published by Acorn Press
An imprint of Bible Society Australia
ACN 148 058 306 | Charity licence 19 000 528
GPO Box 4161
Sydney NSW 2001
Australia
www.acornpress.net.au | www.biblesociety.org.au

© Roland Ashby, Chris Mulherin, John Pilbrow and Stephen Ames 2024.
All rights reserved.

ISBN 978-0-647-53344-4

First published by Morning Star Publishing in 2018,
ISBN 978-0-648-45370-3

Roland Ashby, Chris Mulherin, John Pilbrow and Stephen Ames assert their right under section 193 of the *Copyright Act 1968* (Cth) to be identified as the author of this work.

Unless otherwise stated, all Scripture quotations are from the New Revised Standard Version Bible, copyright © 1989 National Council of Churches of Christ in the United States of America. Used by permission. All rights reserved worldwide.

Apart from any fair dealing for the purposes of private study, research, criticism or review, no part of this work may be reproduced by electronic or other means without the permission of the publisher.

All enquiries should be made to ISCAST: www.iscast.org

 A catalogue record for this book is available from the National Library of Australia

Cover and text design and layout by John Healy

Table of Contents

Preface	7
Foreword	9
List of Authors	11
Historical Perspectives	15
1. Christianity: The womb of Western science	17
2. The modern idea of "science" and "religion" led to "hardening of the edges"	25
3. Science, faith "intimately connected": An interview with Peter Harrison	27
Present Currents: Atheism, Secularism and the Spirits of the Age	33
4. Challenging the prevailing secularist narrative	35
5. Questions for Phillip Adams on his rejection of faith	40
6. Lawrence Krauss takes atheism to a new low	44
7. Thank God for "The Rise of Atheism" convention	49
8. The New Atheists and fundamentalists need to move on	53
9. The "spirit of the age" is out of touch with God's call	59
10. Alain de Botton's *Religion for Atheists* and the varieties of non-religious belief	63
Good, Evil and God	67
11. Genetics, evolution, cancer, suffering and God: An interview with Graeme Finlay	69
12. Challenging Stephen Fry's diatribe against God	75
13. Wrestling with evil: Can a "reckless" creator God be good?	80
14. After disaster, allow space for hard questions	85
15. Droughts and flooding rains: Disasters are part of the planet's functioning	88
16. Natural disasters not a result of the fall says Bob White	91
17. Why does God allow suffering?	96

Future Challenges: Technology, Robots, and Caring for Creation — 101

18. Can the computer be a substitute for humanity? — 103
19. Do robots spell the end for the human race? — 107
20. God is not just the God of human beings — 110
21. God's sustaining power at the heart of life — 114
22. Climate science and Christian faith share much in common — 117
23. Why a Christian response to climate change? — 120
24. Why climate change is real and human-induced — 125
25. It is not only humans that matter to God — 128
26. John Houghton: Lessons from a leading climate scientist and Christian — 130
27. Michael Northcott: Our devotion to idols is killing the planet — 133
28. Is Earth 2.0 really "bad news for God"? — 138

Minds, Brains and Spirituality — 143

29. Meditation, mindfulness and the brain — 145
30. Psychical research a wonderful gift to religion — 151
31. Religious sense at the heart of self-perception — 154

The Human Story — 159

32. Graeme Clark: A lifelong mission to bring hearing to the deaf — 161
33. Jürgen Moltmann: How "Christ the bridge" led me from science to theology — 163
34. John Lennox: Christian apologist in an age of doubt — 175
35. Antony Flew: The philosopher who flew the atheists' nest — 179
36. Francis Collins: Work on human genome strengthened his faith — 183
37. Interviewing Alister McGrath: Science is helpful, but only part of the picture — 187
38. Allan Day: A scientist's quest for what it means to be human — 196
39. Stephen O'Leary: Using science as a tool in the Lord's service — 201

Science and Christian Faith: The State of the Marriage — 207

40. The Christian conflict with science is dead — 209
41. Is life without purpose, and religion irrelevant? — 214
42. A "more subtle science" reveals divine reality — 218

Table of Contents

43.	A brilliant exploration of the science–religion debate	222
44.	Christianity defended with logic and vigour	224
45.	Christianity, science and rumours of divorce	226
46.	Science as ideology betrays its purpose	231
47.	Science is a deeply religious activity	236
48.	Science cannot be an exclusive guide to reality	241
49.	Science is "entirely within God's purposes"	246
50.	Science is wonderful, but it's not the only game in town	251
51.	There is more science–faith overlap than ever	256

Design, the Universe, Fine Tuning, Reality and Metaphysics — 261

52.	A purposeful universe	263
53.	A universe from nothing? What could it mean?	268
54.	Faith, hope and quarks: The search for God	272
55.	Fine tuning: Compelling evidence for God?	277
56.	Does the "God Particle" bring us closer to God?	281
57.	God a "combination of love and mathematics"	284
58.	Gravitational waves discovery opens new way of looking into the universe	286
59.	How science can help us understand prayer	292
60.	In the "great silence" can we be sure we are alone?	296
61.	Quantum uncertainty and the action of God	299
62.	Intelligent Design theory scientifically and theologically flawed	302

Darwin, Biology and Genetics — 305

63.	Adapting to the impact of Darwin	307
64.	Anglican encouragement led to Darwin's science	312
65.	Divine activity through evolution long accepted	315
66.	Genetics: One way God speaks to us	318
67.	Simon Conway Morris: Human race not the result of "dumb luck"	322

Afterword: Faith in a World of Science — 325

Preface

What sort of God would create such an unimaginably vast and life-giving universe? What sort of God would dare to create human beings with such extraordinary powers of creation—powers that can be used for good or for ill? What sort of God would create a world of such beauty and good, yet also a world of earthquakes and tsunamis, cancer and genetic disorders?

In the words of one of this book's authors, the Christian God is the loving, "reckless" creator God who, like the father of the prodigal, risks himself for the sake of relationship. So, this book is about the wonders of a universe that has given birth to life, and it is about the challenges facing human beings, made as co-creators in the image of the "reckless" creator God.

How, then, does the scientific discovery of the wonders of the universe relate to Christian understandings of God the creator?

We believe there is no fundamental conflict between a scientific outlook and the Christian faith. Indeed, we contend there is a demonstrable coherence between the accounts of the world offered in Scripture and Christian theology, and the accounts of mainstream science. While theology and Scripture are concerned with *meaning* and the purposes of God (sometimes characterised as the "why?" questions) science is concerned with *mechanisms* and the particles of the universe (the "how?" questions).

So, we are united in asserting that the Bible is not a scientific textbook and that a fuller understanding of the creation in which we find ourselves is encountered through the methods of the many branches of science. However, we make no claim that scientific knowledge as we understand it today is exhaustive or even entirely accurate; science, like theology, is a continuing conversation.

We believe that "all truth is God's truth" and if there appears to be a conflict between scientific claims and Christian faith then we are challenged to re-examine our understanding of either one or the other, or even both, to see if perhaps better understandings are possible.

It is without doubt that the discoveries of modern science have changed our world and our lives in ways our predecessors could not possibly have imagined. So it is imperative that Christians are equipped to think about these developments maturely. While this volume does not claim to provide all the answers, these chapters provide important insights as well as encouragement to pursue the issues.

A Reckless God?

Contributors include distinguished scholars such as Alister McGrath, a globally recognised thinker in matters of theology and science. Another notable contributor, the renowned German theologian Jürgen Moltmann, is one of a small number of contemporary theologians who have wrestled with the theological implications of modern science. His chapter not only explores the relationship between science and faith but also tells his personal journey that began in a POW camp in the UK after the First World War. Another author, Peter Harrison, is an eminent Australian scholar known for his historical work on the emergence of science in Christendom.

This ISCAST Nexus book is published by ISCAST—Christians in Science and Technology in conjunction with *The Melbourne Anglican*. The book largely consists of essays that have appeared over recent years in the Faith–Science Interface series published in *The Melbourne Anglican*. The series is intended to stimulate and sustain a conversation about the relationship between science and Christian faith—a conversation that is, necessarily, ongoing.

ISCAST is an Australian network of people, from students to distinguished academics exploring the interface of science, technology and Christian faith. More details about ISCAST can be found at www.iscast.org.

We commend this volume as an Australian contribution to the extensive literature now available.

    Roland Ashby
    Chris Mulherin
    John Pilbrow
    Stephen Ames

    Melbourne, Australia
    November 2018

Foreword

By Jennifer Wiseman

Jennifer Wiseman is an astrophysicist, speaker, and author. She is a Fellow of the American Scientific Affiliation, a network of Christians in science, and she is a Distinguished Fellow of ISCAST—Christians in Science and Technology. She directs the Dialogue on Science, Ethics, and Religion for the American Association for the Advancement of Science.

Science and technology impact nearly every aspect of modern daily life, whether or not we realise it. Communications, health care, agriculture, entertainment, transportation, exploration and education are all enabled, shaped and even guided by advancements in technology and a scientific understanding of the natural world.

Likewise, for the committed Christian, faith in God, respect for truths and wisdom revealed in Scripture, and an interactive relationship with God frame every aspect of life's values, choices and purpose. Are these two frameworks—scientific understanding and Christian belief—complementary to each other? Are they conflicting? Can one live within one framework but reject or ignore the other? Can they inform or enrich one another? These questions and related assumptions, whether voiced or unspoken, are reflected in modern media, education and public discourse. They even impact public policy decisions, as ethical boundaries for the uses of technology are becoming more complex and involve value judgements outside the purview of science alone by people looking through their own broader worldview lenses.

Thus while the relationship between science and religious faith has been contemplated since the birth of modern science, this book offers a fresh look at that interface in the context of current discovery, scientific advances and global ethical challenges. The authors and voices in this timely collection of essays and conversations reflect on several facets of the contemporary and future conversation between science, technology and Christian faith.

Several realms of scientific advancement are particularly fertile for such dialogue. Gene mapping, for example, is showing in exquisite detail the interrelatedness of all life on earth. It is also opening up ethical challenges never seen before, with the advent of the ability to edit human genes, and therefore possibly humanity itself. Robotics and artificial intelligence are blurring the line between human and machine. Personal electronics and electronic communications are changing the way humans relate to one another in profound ways not yet fully grasped. Neuroscience is unveiling the connections between brain, mind and behaviour.

These all point to fundamental questions: What does it mean to be human? Are we completely defined by our genome? Are we only our brains? Can machines and computers replicate and improve upon flesh and blood?

This conversation is enriched with the acknowledgment that Christian faith is primarily about relationships in a way that science is not. Relationships between God and people, between persons, and between people and nature are central to the teachings of Scripture. Therefore advances in science and technology, rightly applied, can be used within the commitments of faith to understand human relationships to nature and to serve others with great efficacy. Understanding the causes and effects of global climate change, for example, can help faith communities advocate broadly for lowering greenhouse gas production while also helping people most vulnerable to the environmental impacts of climate change. Likewise, advances in medical and agricultural technology can be translated via faith communities into actual help and blessing for people in need around the world. Knowledge of ecology and animal behaviour combined with human compassion can lead to better protections for animals and their habitats around the world, whether in the wilderness or on the farm, in labs or in the home.

Looking out beyond this planet alone, our studies of distant galaxies are revealing a very real connection between life on earth and the billions-of-years' evolution of the universe. Our own bodies contain atoms formed in stars! And now thousands of planets—and in the future, potentially life forms—are being discovered outside of our own solar system. Sharing such knowledge and discovery with others can lift spirits and sights to the broader universe within which we are all connected. For believers, these discoveries advanced by science can also elicit praise and reverence for the one revered as creator and sustainer of it all.

The contributors to this book provide many perspectives of how the forward march of science and technological advancement mesh, in their lives, with their faith in a loving, personal God, and with related service to other inhabitants of planet Earth. May their "conversations with science" deepen our own reflections on the broader questions of life and meaning in this age of exciting scientific discovery.

List of Authors

Roland Ashby has been the Editor of *The Melbourne Anglican* and Director of Anglican Media Melbourne for the Anglican Diocese of Melbourne since 1995. He holds an MA in theology, is a Benedictine Oblate, and is a member of the World Community for Christian Meditation. Roland is author of *A Faith to Live By: Conversations About Faith with 25 of the World's Leading Spiritual Teachers* (Morning Star, 2012), and editor of *A Faith to Live By* Vol 2 (Morning Star, 2018) and *Heroes of the Faith: Fifty-five Men and Women Whose Lives have Proclaimed Christ and Inspired the Faith of Others* (Garratt, 2015).

The Rev. Dr Stephen Ames is an Anglican priest and a canon of St Paul's Cathedral, with doctorates in both physics and the history and philosophy of science. He lectures in God and the Natural Sciences at the University of Melbourne and is organiser of the annual Science Week at the Cathedral at St Paul's Cathedral, Melbourne. He is also a Fellow of ISCAST.

Scott Buchanan is studying theology at Ridley College, Melbourne, and is a social worker in community mental health. He has a blog at scottlbuchanan.wordpress.com.

Dr Tom Butler is the former Bishop of Southwark, in the UK.

Dr Jonathan Clarke is a geologist and a Fellow and Director of ISCAST.

Dr Shane Clifton is Dean of Theology at Alphacrucis College, Australia.

Dr Denise Cooper-Clarke is a voluntary researcher with ETHOS (the Evangelical Alliance Centre for Christianity and Society), an occasional Adjunct Lecturer in Ethics at Ridley College, Melbourne and a Fellow of ISCAST.

Madeleine Davies is the deputy news editor and a reporter for *The Church Times* in the UK, and one of the paper's features editors.

Dr Allan Day was Emeritus Professor of Physiology, the University of Melbourne and a Senior Academic Fellow, Ridley College. He was also a Fellow of ISCAST. He died in 2013.

A Reckless God?

Dr Alan Gijsbers is head of the Addiction Medicine Service at the Royal Melbourne Hospital and is the Medical Director of the Substance Withdrawal Unit at the Melbourne Clinic, Richmond. He is also President of ISCAST.

Dr Colin Goodwin is a retired Anglican priest and lecturer.

Dr David Griggs is Adjunct Professor of Sustainable Development at Monash University and from 2007–2015 was Director of the Monash Sustainability Institute.

Dr Peter Harrison FAHA is an Australian Laureate Fellow and Director of the Institute for Advanced Studies in the Humanities at the University of Queensland. Previously he was the Idreos Professor of Science and Religion and Director of the Ian Ramsey Centre at the University of Oxford. Author of over 100 articles and book chapters, his six books include *The Territories of Science and Religion* (University of Chicago Press, 2015). He is a Fellow of ISCAST.

The Rev. Murray Hogg trained in engineering, is a Fellow of ISCAST and the Pastor of East Camberwell Baptist Church.

Dr Rodney Holder is Emeritus Course Director of The Faraday Institute for Science and Religion, Cambridge, UK, and is a Fellow Commoner of St Edmund's College, Cambridge. He read mathematics at Trinity College, Cambridge, and was awarded a DPhil in astrophysics and later a theology degree from Oxford University. He is a priest in the Church of England. His books include *Big Bang, Big God: A Universe Designed for Life?* (Lion Hudson, 2013).

The Rev. Dr Mark Lindsay is Joan F. W. Munro Professor of Historical Theology, Trinity College Theological School, Melbourne.

Dr Donald MacKay was Professor of Communication and Neuroscience, University of Keele, UK. He died in 1987.

The Rev. Dr Alister McGrath is Idreos Professor of Science and Religion at Oxford University. Previously he was Professor of Theology, Ministry and Education King's College London, Professor of Historical Theology at Oxford and Principal of Wycliffe Hall, Oxford.

List of Authors

Dr Tom McLeish is Professor of Physics at Durham University, UK, a Fellow of the Royal Society and author of *Faith and Wisdom in Science*. He is also a Distinguished Fellow of ISCAST.

Dr Jürgen Moltmann is one of the most widely read theologians of the last 60 years. He is Professor Emeritus of Systematic Theology at the University of Tübingen, Germany.

The Rev. Dr Chris Mulherin trained in engineering and philosophy and is an Anglican minister. He is Executive Director of ISCAST and teaches philosophy.

The Rev. Dr Steven Ogden is Principal of St Francis' Theological College, Brisbane, Australia. He is the author of *The Church, Authority, and Foucault— Imagining the Church as an Open Space of Freedom* (Routledge, 2017).

Fr Daniel O'Leary is a Roman Catholic priest, author, teacher and retreat leader in the Diocese of Leeds, UK.

Dr Stephen O'Leary is an ear, nose and throat surgeon and holds the William Gibson Chair of Otolaryngology at The University of Melbourne. He obtained his PhD under the mentorship of the inventor of the bionic ear, Professor Graeme Clark.

Dr Mick Pope is a meteorologist and head of the ETHOS Environment Think Tank for ETHOS, the Evangelical Alliance Centre for Christianity and Society in Australia.

Dr John Pilbrow is Emeritus Professor of Physics at Monash University and formerly President of The Australian Institute of Physics. He is former President and now an Honorary Life Fellow of ISCAST.

The Rev. Dr Duncan Reid is Head of Religious Education at Camberwell Girls' Grammar School, Melbourne, Australia.

Fr Paul Richardson is a Roman Catholic priest and a former Anglican Bishop of Wangaratta, Australia, and former Assistant Bishop in the Anglican Diocese of Newcastle in the UK.

A Reckless God?

The Rev. Dr Barry Rogers, is a secular priest, working as a consulting psychologist in private practice, including mentoring for people in ministry and business. He is also a Consultant Psychologist with the Department of Theological Education for the Anglican Diocese of Melbourne's assessment process for those seeking ordination.

Dr Mark Vernon is a specialist writer on science and religion in the UK. His latest book *God: The Big Questions* is published by Quercus.

The Rev. Dr David Wilkinson is Principal of St John's College, the University of Durham, and formerly an astrophysicist.

Dr Jennifer Wiseman is an astrophysicist, speaker, and author. She is a Fellow of the American Scientific Affiliation, a network of Christians in science, and she is a Distinguished Fellow of ISCAST. She directs the Dialogue on Science, Ethics, and Religion for the American Association for the Advancement of Science.

Dr Andrew Wood is Professor of Biophysics at Swinburne University of Technology. He is a Fellow of ISCAST.

The Rev. Dr Mark Worthing is a Lutheran pastor and theologian, with doctorates in theology and the history and philosophy of science. He is a Fellow of ISCAST and formerly Senior Researcher at the Australian Lutheran Institute for Theology and Ethics, Australian Lutheran College, Adelaide. He is now Pastor of Immanuel Lutheran Church, North Adelaide.

Dr David Young is a Principal Fellow in the Department of Zoology at the University of Melbourne and author of *The Discovery of Evolution* (Second Edition, Natural History Museum & Cambridge University Press, 2007).

Historical Perspectives

The first three chapters in this book concern the historical context that informs discussion of science and religion in the West. All three pieces relate to Peter Harrison's work on the history of the relationship between science and religion, particularly Christianity, in the West during "the rise of science." Harrison, who heads the Institute for Advanced Studies in the Humanities at the University of Queensland, is a recognised authority in the field. The first chapter is by Harrison, the second is a review of his *The Territories of Science and Religion*, and the third is an interview with Harrison.

1. Christianity: The womb of Western science

By Peter Harrison

> Those who magnify controversies about science and religion, projecting conflict back into historical time, perpetuate an historical myth to which no historian of science would subscribe.

It is often assumed that the relationship between Christianity and science has been a long and troubled one. Such assumptions draw support from a variety of sources.

There are contemporary controversies about evolution and creation, for example, which are thought to typify past relations between science and religion. This view is reinforced by popular accounts of such historical episodes as the condemnation of Galileo, which saw the Catholic Church censure Galileo for teaching that the earth revolved around the sun.

Adding further credence to this view of history are a few recent outspoken critics of religion who vociferously contend that religious faith is incompatible with a scientific outlook, and that this has always been the case.

In spite of this widespread view on the historical relations between science and religion, historians of science have long known that religious factors played a significantly positive role in the emergence and persistence of modern science in the West. Not only were many of the key figures in the rise of science individuals with sincere religious commitments, but the new approaches to nature that they pioneered were underpinned in various ways by religious assumptions.

The idea, first proposed in the seventeenth century, that nature was governed by mathematical laws, was directly informed by theological considerations. The move towards offering mechanical explanations in physics also owed much to a particular religious perspective.

The adoption of more literal approaches to the interpretation of the Bible, usually assumed to have been an impediment to science, also had an important, if indirect, role in these developments, promoting a non-symbolic and utilitarian understanding of the natural world which was conducive to the scientific approach.

Finally, religion also provided social sanctions for the pursuit of science, ensuring that it would become a permanent and central feature of the culture of the modern West.

Historical Perspectives

God and the laws of nature

The remarkable scientific achievements of Isaac Newton (1642–1727) were neatly summarised in poet Alexander Pope's famous couplet: "Nature and Nature's laws lay hid in night/God said: 'Let Newton be!' and all was light." Pope's laudatory verse makes the point, known to every student of physics, that Newton discovered some of the fundamental laws of nature.

What is often less appreciated, however, is that part of the novelty of Newton's achievement lay in his conviction that there were laws of nature there in the first place, awaiting discovery. What, then, are laws of nature, and where do they come from? During the Middle Ages, natural laws were understood to be moral laws that had been established by God. The injunction "Thou shalt not kill" was an example of one such law, assumed to be a universal rule that all civilised societies would observe. However, there was no corresponding notion of universal laws in the natural realm.

This idea—laws of nature in the scientific sense—was an innovation of the seventeenth century and was a consequence of the extension of God's legislative moral power to the physical world. One of the pioneers of this new understanding of laws of nature was the French philosopher and scientist Rene Descartes (1596–1650), who wrote that "God alone is the author of all the motions in the world."

This was a radical claim for its time, for it challenged the prevailing view, inherited from the Greek philosopher Aristotle (384–322 BC), that the behaviours and interactions of material objects were governed by their internal properties. According to the Aristotelian worldview, which had held sway throughout the Middle Ages, nature had enjoyed a considerable degree of autonomy.

In the new science, however, natural objects were stripped of inherent properties and God assumed direct control of their interactions. In much the same way that the Deity had instituted moral rules, he was now thought to enact laws which governed the natural world. Robert Boyle, the father of modern chemistry and author of the eponymous law, observed that God's creation operates according to fixed laws "which He alone at first Establish'd." God's authorship of the laws of nature guaranteed their universality and unchanging nature. Descartes thus argued that because these laws had their source in an eternal and unchanging God, the laws of nature must themselves be eternal and unchanging.

Descartes also set out a law of the conservation of motion, again arguing for it on the basis of God's immutability. This idea that nature was governed by constant and immutable principles was an important precondition for experimental science.

The mathematician Isaac Barrow, who was Isaac Newton's predecessor in the famous Lucasian Chair of Mathematics at Cambridge, suggested that the only reason for having confidence that repeated experiments will yield general principles that hold true, is that we can be assured that the laws of nature that God has instituted are constant. We have no reason to believe, he wrote, "that Nature is inconstant," for that would imply that "the great Author of the universe is unlike himself."

Mathematics and cosmic order

Closely linked to the new idea that nature was governed by universal and immutable laws was the introduction of mathematical explanation to the natural sciences. Again, this was something quite new.

While medieval thinkers had worked with mathematical equations in the spheres of optics, astronomy and kinematics, they had tended to regard these disciplines as something less than true science. The status of these "mixed mathematical sciences," as they were known, was a consequence of the Aristotelian understanding of mathematics and its relation to the other sciences.

One of the perennial questions in the philosophy of mathematics concerns the status of mathematical truths: Are they human constructions, or are they eternal truths that are embedded in reality? Plato had held the first position, Aristotle had adopted the second, and it was Aristotle's view that tended to prevail throughout the Middle Ages.

If mathematics was primarily a product of the human mind, it could be argued that mathematics did not necessarily provide a true description of reality. It might be allowed, however, that mathematical models, although not ultimately true, nonetheless provided the basis for accurate predictions.

Hence, mathematical astronomy, while regarded as falling short of offering a true account of the nature of heavenly bodies and the causes of their motions, was regarded as useful because it made it possible to predict their positions. Mathematical models were thus thought of as *useful fictions*.

It was a difference of opinion on this question that led to Galileo's confrontation with the Inquisition. Galileo had wanted to insist that the sun-centred Copernican model system was more than a helpful mathematical device—it was an accurate physical description. Thus, not only did Galileo champion a new astronomical model, he also held to a new model of astronomy. The Catholic Church, for its part, supported the prevailing view. With the benefit of hindsight, we might judge its decision to have been unwise, but it was consistent with the scientific consensus of the time.

As was the case for laws of nature, the idea that mathematical relations were real had a theological justification. Individuals such as Galileo, Johannes Kepler, Rene Descartes and Isaac Newton were convinced that mathematical truths were not the products of human minds, but of the divine mind. God was the source of mathematical relations that were evident in the new laws of the universe. Like the Bible, the "book of nature" had also been written by God and, as Galileo was to insist, this book was "written in the language of mathematics."

Other scientists shared this view. Johannes Kepler, who discovered the laws of planetary motion, argued that God had used mathematical archetypes in his creation of the cosmos. Because of this, he wrote, the old Aristotelian prejudice against the mathematisation of nature was to be rejected:

> The reason why the mathematicals are the cause of natural things (a theory which Aristotle carped at in so many places) is that God the creator had mathematicals with him as archetypes from eternity in their simplest divine state of abstraction.

Descartes even claimed that God had created the laws of logic and mathematics, maintaining that the equation $2 + 2 = 4$ was true only because God had so willed it. In support of the idea that God was a mathematician, Descartes quoted the biblical verse: "thou hast ordered all things in measure and number and weight" (Wisdom of Solomon 11.20). Newton subsequently described the cosmos as inhabited by an "infinite and omnipresent spirit" in which matter was moved by "mathematical laws."

Identifying God as the author of mathematics was thus a crucial step in asserting the reality of mathematical relations, and it was this development which enabled the subsequent application of mathematics to the subject matter of physics. Combined with the idea of a divine legislator, this insight produced the modern view that nature is governed by mathematical laws.

Transitioning from medieval to early modern worldviews

In the medieval understanding of nature, indebted as it was to Aristotelian science, the activities of material things were governed by their internal properties, usually understood in terms of such qualities as heat, cold, moisture and dryness. Aristotle had also imagined that objects had natural tendencies, or "final causes." Nature, in this scheme of things, was self-organising and was conceptualised as analogous in many respects to a living thing.

The seventeenth century saw the revival of an alternative conception of nature—the atomic or "corpuscular" theory that had been championed by the ancient Epicureans. According to this view, matter was made of minute particles that were more or less qualitatively identical. These particles were able to combine in various ways to form macroscopic matter, and the operations of nature were explained by the interactions of the invisible particles.

Such qualities as heat and cold could thus be accounted for in terms of the motions of minute particles, rather than being considered as inherent qualities of particular kinds of substances. The base units of matter were also imagined to fall within the explanatory range of the laws of nature. In much the same way that the movements of the heavenly bodies were expressed in terms of mathematical laws, so too were the movements of the minute corpuscles of matter.

Nature was also increasingly understood as analogous to a machine, rather than a living organism. For this reason, the new science was often referred to as "the mechanical philosophy." Just as machines were built and designed by human agents, the world was believed to exhibit evidence of design by a divine agent (Aristotle, although a theist, had thought the world eternal and hence uncreated).

Accordingly, the idea of "final cause" underwent a major change, and was now typically identified with God's purposes or designs. The idea that nature bore witness to the designs of a deity became a powerful justification for the pursuit of natural science. Indeed, while it is rightly thought that seventeenth and eighteenth-century natural science lent rational support to Christian belief, it is equally true that the natural sciences gained in social legitimacy because they were perceived to be theologically useful.

In each of these seventeenth-century developments—the emergence of natural laws, the mathematisation of nature, the new mechanistic and atomic understanding of matter—God was imagined to be more intimately involved in nature than he had been in the medieval world picture. Indeed, this was the explicit intention of some of the principal agents of the scientific revolution, who argued that their new views of nature were more genuinely Christian than the supposedly "pagan" science of Aristotle.

There is an element of truth in the common assumption that the transition from medieval to early modern worldviews that took place over the course of the sixteenth and seventeenth centuries inevitably resulted in a secularisation of the cosmos. Yet, many of the leading figures in the scientific revolution imagined themselves to be champions of a science that was more compatible with Christianity than the medieval ideas about the natural world that they replaced.

A renewed emphasis on the sovereignty of God is evident in all of these developments. This paralleled a movement simultaneously unfolding in the theological sphere with regard to the doctrine of justification. In an argument analogous to that which stripped natural bodies of their inherent causal virtues, Protestant reformers insisted that inherent human virtues were not causally efficacious in bringing about justification.

The whole initiative for the process of salvation lay with God, whose eternal decree determined who would be justified. Developments in both spheres, the theological and the natural, are linked to renewed assertions of the sovereignty of God, and to the understanding of the exercise of that sovereignty in terms of eternal and unchangeable laws.

The book of nature and the book of Scripture

If physical objects were stripped of their intrinsic qualities in the new conception of nature, they were also denied any function as natural symbols. Throughout the Middle Ages, features of the natural world had symbolised theological and moral truths.

It is often thought that the rise of modern science caused the death of this rich symbolic world. In fact, it was the demise of symbolism, promoted in various ways by Renaissance humanists and Protestant reformers, which helped make space for the new science. Galileo's reference to "the two books" reminds us of important connections between the two modes of God's communication, nature and Scripture. For much of the Middle Ages the "books" of nature and Scripture had been read together as part of a unified interpretive endeavour.

The link between the two books was provided by allegorical interpretation. Augustine and Thomas Aquinas had both argued that the literal meaning of Scripture is established by identifying the objects to which the words refer. The allegorical sense, however, was to do with the meanings of those objects. For the allegorical reader, the words of Scripture directed the senses to a natural world in which objects bore rich theological and moral

Christianity: The womb of Western science

meanings. Allegory, in short, was a technique for reading the world through Scripture.

Owing to the combined efforts of Renaissance humanists and Protestant reformers, the sixteenth century saw a decline in allegory and a renewed emphasis on the literal sense of the Bible. In his typically colourful prose Luther observed that allegorical readings were for "weak minds" and "idle men." The Reformation helped precipitate the collapse of allegory by casting suspicion on visual representation and symbolism generally.

An unintended consequence of these developments was that natural objects were no longer read for their theological and moral meanings. This in turn raised acute questions about how nature was to be understood, and its purpose in the divine economy. It is in this sense that the world was indeed desacralised at this time, although this was not a consequence of science, but rather a shift in Western sensibilities that helped make science possible.

One response to new questions about the intelligibility of nature was reconfiguration of nature in terms of mathematical relations. This solution is evident in Galileo's new deployment of the "book of nature" metaphor. The decline of allegorical readings of nature thus opened the way for mathematical readings.

Equally importantly, nature was increasingly regarded as a realm to be exploited for material benefit rather than for moral and theological edification. This led to attempts to master the natural world, which were themselves motivated by new literal readings of the creation narratives, and in particular the Genesis injunctions to subdue the earth and exercise dominion over it (Gen 1:28; 9:2). Religious considerations thus provided important sanctions for the utilitarian orientation of modern science.

Francis Bacon and his successors in the Royal Society, for example, clearly saw themselves as attempting to regain the dominion over nature which Adam had forfeited as a consequence of his disobedience. As Bacon expressed it:

> For man by the fall fell at the same time from this state of innocency and from his dominion over creation. Both of these losses however can even in this life be in some part repaired; the former by religion and faith, the latter by arts and sciences.

Scientific activity thus came to be regarded as an integral part of a redemptive process. This more active engagement with the natural world was still pursued from theological motives, but clearly these were quite different from those of medieval allegorists.

Theological roots of science?

Could modern science have arisen outside the theological matrix of Western Christendom? It is difficult to say. What can be said for certain is that it did arise in that environment, and that theological ideas underpinned some of its central assumptions. Those who argue for the incompatibility of science and religion will draw little comfort from history.

What the historical record also suggests is that insofar as modern science posits natural laws and presupposes the constancy of nature, it invokes an implicit theology. Most important of all, perhaps, religious considerations provided vital sanctions for the pursuit of scientific knowledge and, arguably, it is these that account for the positive attitudes to science which have led to the high status of science in the modern West.

This is not to deny that there have been those in the past who have opposed certain scientific views on religious grounds. This has been especially the case since the advent of Darwinism, which met with a mixed reception in religious circles. It is often forgotten, however, that Darwinism met with a mixed reaction in scientific circles, too.

Those who have magnified more recent controversies about the relations of science and religion, and who have projected them back into historical time, simply perpetuate a historical myth. The myth of a perennial conflict between science and religion is one to which no historian of science would subscribe.

> This chapter is a slightly edited version of an article which first appeared online at ABC Religion and Ethics.

2. The modern idea of "science" and "religion" led to "hardening of the edges"

By Duncan Reid

A review of *The Territories of Science and Religion*, by Peter Harrison (University of Chicago Press, 2015).

Since the renewed collaboration between theologians and scientists over recent decades, there has been no shortage of scholarly refutations of the notorious late nineteenth-century mythology of "warfare" between science and religion. The main promulgator of this myth was Andrew Dickson White, whose *History of the Warfare of Science with Theology in Christendom* was published in 1897. White's argument has been roundly discredited, generally by way of point by point argument. Despite these efforts, White's warfare fantasy continues to have a massive hold on the popular imagination, with Galileo and Darwin often cited as heroes of science struggling in the face of religious obscurantism. *The Territories of Science and Religion* by University of Queensland researcher Peter Harrison acknowledges the efforts to respond to this misperception, but proposes a far more economical and elegant dismissal of the warfare myth, as simply anachronistic. Current popular discussion, he suggests, is "often oblivious to the problematic nature of the categories in question" (p. 6).

Harrison's argument—in just 200 pages, but supported by a further 100 pages of footnotes, index and bibliography—is simply that the categories in question, namely "science" and "religion," as we commonly use the terms today, are essentially modern concepts. The idea of religion, as a set of propositionally formulated truth claims, dates only to the seventeenth century. Before that, "religion" denoted an inner attitude or approach to life, something far closer to what we would now call piety or spirituality. Harrison illustrates this with reference to linguistic changes in, among other things, successive translations of Calvin's *Institutio Christianae Religionis*, where "institution of Christian religion" (suggesting the origins of an inner disposition) is reified into *the* institution of *the* Christian religion (the definite articles functioning to define *this* form of religion over against other possible forms).

A similar hardening of the edges is noted in moves from speaking of "true religion" (a phrase some of us will remember from the Book of Common

Prayer) to "*the* true religion." In both cases, "the content of catechisms that has once been understood as techniques for instilling an interior piety now came to be thought of as encapsulating the essence of some objective thing—religion" (p. 84). The tendency to speak in these more definite terms aided the division of early modern Europe into different "religions," Catholic and Protestant, and later gave impetus to the European colonial project in its attempt to categorise all "other religions."

Science underwent a similar transformation, but even later than religion, from a cultivated habit of thinking about nature—a natural philosophy as it was understood even as late as the mid-nineteenth century—reflecting again a certain "mental disposition" (p. 69), to a particular clearly defined "method" carried out by a clearly credentialed type of practitioner, the "scientist." This Harrison sees as a political move "to establish a scientific status for natural history, to rid the discipline of women, amateurs, and parsons, and to place a secular science into the centre of cultural life in Victorian England" (p. 183). The success of this program is demonstrated by an accompanying graph of the decreasing overall membership of the Royal Society, and the decreasing proportion of Anglican clergy among its membership, as admission rules were tightened in the second half of the nineteenth century.

Harrison's book opens with an invitation to consider a map from 400 years ago: territories may have familiar names, but it would be a category mistake to imagine that their inhabitants saw themselves or organised themselves then in ways we would expect in the same territories today. So also, argues Harrison, with the seemingly familiar categories "science" and "religion": their territories of scope and connotation have shifted beyond recognition. Harrison also warns, however, against drawing any hasty conclusions about a possible congruence of science and religion: here again any uncritical or ahistorical use of the categories can be misleading.

This is a book that deserves the attention of anyone with an interest in science or religion, or the relationship between them; or indeed, the consequences of this relationship.

3. Science, faith "intimately connected": An interview with Peter Harrison

By Chris Mulherin

Australian historian and philosopher Professor Peter Harrison spoke to Chris Mulherin on the history of the relationship between science and religion in the West. Contrary to popular belief, he says, science and religion were understood as intimately connected until recent times.

Peter Harrison says we've been duped into believing in the myth of a perennial conflict between science and Christianity. No, Harrison doesn't use the word "duped," because he is quiet-spoken, choosing his words with care, and wary of overstating his case.

Harrison returned home to the University of Queensland a number of years ago after spending time at Yale University and then as Idreos Professor of Science and Religion at Oxford University, where he also headed the Ian Ramsay Centre for Science and Religion. His best known book, *The Territories of Science and Religion*, is based on the prestigious Gifford Lectures he gave at the University of Edinburgh in 2011.

Harrison is now an Australian Laureate Fellow and Director of the Institute for Advanced Studies in the Humanities. He is also a Fellow of ISCAST—Christians in Science and Technology, and he was in Melbourne recently to give the annual ISCAST Allan Day memorial lecture at Ridley College. He also presented a public lecture at the University of Melbourne, and a staff and postgraduate seminar at the University of Divinity.

Our conversation ranged from Galileo, Darwin's dogs (a bulldog and a Rottweiler), the Western roots of science, to the question of whether science, come of age, makes religion redundant.

The conflict myth

We started talking about the so-called conflict between science and faith, with Harrison affirming that conflict is the way the relationship is most commonly understood—both outside the church, and also from anti-evolution, "creation-science" movements within the Christian church. In Islamic societies too science is often seen as being in conflict with religion.

Historically, however, the relationship has been a mixed one, sometimes positive, sometimes not so. But overall, Harrison says, "the relations have been much more positive on balance than negative." In fact, historians of science, including those who have no interest in being apologists for religion, are clear that the conflict story is a mistaken view of the past: "if you look at what historians of the science–religion relationship have said, probably over the past fifty years, it's a puzzlement about the persistence of this myth."

The Galileo affair

The so-called Galileo affair is, perhaps, the most celebrated example of acrimonious relations between faith and science. The myth pits the truth-seeking "scientist" of the early seventeenth century against the anti-science Roman Catholic Church. However, the truth is more complex: on the one hand "the scientific consensus at the time was firmly against the view that Galileo was proposing" and, on the other hand, the response of the Church was "in no way typical of the Catholic attitude to science." Added to that is a background of personality clashes and political intrigue in Rome.

For the Catholic Church, as a long-time supporter of science, there were good scientific reasons that Galileo's hypothesis of a moving earth (as opposed to the stationary earth of the long-accepted Aristotelian view of the universe) was deeply problematic. Harrison says, "the Church did a lot of work to try to get the science right, and to that extent they were essentially supporting what they believed to be the correct scientific consensus. And what that suggests is that there are a number of difficulties with the scientific claims that Galileo was making."

As well as the scientific difficulties of Galileo's position, Catholic support of science needs to be recognised: "no other institution from the late Middle Ages had supported astronomical research to the extent that the Catholic Church did." As well, Harrison cites the medieval universities, started by the Church and the sites where science was conducted: "They were huge supporters of knowledge and learning," he says.

So, despite the fact that Galileo remained a faithful Catholic to the end of his life, the myth of "Science v. Church" in the seventeenth century prevails.

In England too, says Harrison, "from the seventeenth century to the late nineteenth century there is a very strong consensus that brings theology and science together. ... The Church of England is very much behind science; key scientific figures are religious figures, and theology is the medium for the popularisation of science. So, that's why people are interested in science: because it has theological pay-off."

However, in the nineteenth century the conflict hypothesis takes its shape.

Evolution and Darwin's bulldog

In 1859 Charles Darwin published his *Origin of Species* challenging some understandings of human uniqueness. Thomas Henry Huxley vigorously promoted a conflict between science and the church and was known as "Darwin's bulldog" (so, today, New Atheist Richard Dawkins is "Darwin's Rottweiler"). According to Harrison, "Huxley came from a working-class background, wasn't educated at Cambridge or Oxford, and resented the fact that the Anglican establishment had control over scientific positions in Europe. So, he wanted to liberate science from ecclesiastical control."

In this period, a prerequisite for taking degrees at Oxford or Cambridge was signing on to the Anglican articles of faith. So, says Harrison,

> the control of the Church of England extended to the control of university positions. Huxley wanted to break that monopoly and thought that science was the way to do it. And he explicitly wanted to use evolutionary theory as a way of undermining the authority of the Church and setting up a conflict between science and religion.

The construction of the conflict hypothesis

The conflict hypothesis took formal shape in the work of two authors and hence is often known as the Draper-White hypothesis. John William Draper was an English-American chemist who wrote *The History of the Conflict Between Science and Religion* in 1874, and, for the first time, says Harrison, "we get the quite explicit articulation of the idea that there's a long-standing conflict between science and religion historically." Andrew Dickson White was President of Cornell University in the US and "concerned that some of the clergymen where Cornell was based were wanting to resist his attempts to make the university a place where secular science could get up and running." In 1896 White published *A History of the Warfare of Science with Theology in Christendom*. Harrison continues:

> From that time on this idea of a conflict between science and religion became a kind of key theme. There are good stories there about the lone scientific hero battling the forces of religious dogmatism— Galileo the lone scientist, persecuted by the Inquisition ... These

things are like hooks; they get into people's minds and then they're reinforced by contemporary instances of science–religion conflicts such as religious-motivated anti-evolutionism; it all seems to fit together in a coherent view of our history.

Science and faith interdependence?

So, if the conflict story isn't historically right, how does Harrison see the relationship? "If you look at the historical origins of science," he says, "you see a very strong, dependent relationship."

Firstly, many of the key figures of the scientific revolution were motivated by understanding science as the study of God's handiwork. Secondly, early scientists had religious presuppositions for pursuing science (for example, that the natural world is governed by laws of nature which are in effect divine edicts). As well, says Harrison, the experimental approach to the natural world is based on a particular theological understanding of human nature. And finally, early science needs something to give it legitimacy, and religion was a means of providing that context: science was seen as religiously safe, religiously useful, and, for some at least, actually an intrinsically religious activity.

Johannes Kepler, who gave us the laws of planetary motion, wanted to study theology but had a change of mind, saying his work in astronomy—studying God's creation—was just as theological. Robert Boyle (of "Boyle's gas law" fame) was a key figure in the foundation of the Royal Society who saw science as "rational worship of God." And Isaac Newton, author of perhaps the most famous scientific work ever, wrote that when he composed his *Principia Mathematica*, he had in mind providing reasons for people to look at nature and think about God as a consequence.

In short, science and religion were intimately connected in these formative phases, says Harrison, making it possible for science to "become one of the key going concerns in Western culture ... we don't see that at any other time or anywhere else."

Science: Uniquely Western?

But what about science in other places, I wonder? Wasn't China well ahead in the science game? Harrison agrees that China was technologically sophisticated, but says that part of the distinctiveness of the Western rise of science has to do with religious convictions about the universe as an orderly creation of God: "the understanding of how laws of nature operate is a very

Science, faith "intimately connected": An interview with Peter Harrison

distinctive Western conception that we see emerging in the seventeenth century, and China has nothing like that." As well, he says,

> for China there were always other priorities. They understood the importance of technology but to some extent it's clear that they were more interested in human relations: the things that we would associate with humanities rather than sciences. So, they prioritised something that they regarded as more important, and it's quite distinctive of the West that we put so much emphasis on science and what it can deliver.

Has science made religion redundant?

But times have changed; surely science the world over has left behind whatever theological roots it might have had? So, I pose the question on the lips of many secularists today: hasn't science made religion redundant? Harrison's response is enlightening, teasing out the ongoing need of science to work within a bigger frame. I will allow him to have the last word:

> For science to make religion redundant, what would be required is that science do the job that religion once did. We often hear this argument—Richard Dawkins is a famous proponent—that science and religion attempt to offer competing explanations for the same thing. But this involves a huge misunderstanding of what religion is about. Religion is about the ultimate questions: the questions of purpose and the questions to do with the origin of intelligibility—very things that make science possible. Science cannot itself answer these questions; science assumes a degree of intelligibility; for science to work, the world must have an underlying structure, something we can see and make sense of.
>
> The most fundamental question of all is: Why there is a world at all? That's not a scientific question, that's a philosophical or religious question. And once we have the world, why does the world have the particular features that it does? Why does it have an order that we can come to understand? And then, crucially, how is it that our minds, which presumably are the end products of a long process of evolution, can understand the rational structure of the universe?
>
> These questions, I think, are genuinely puzzling unless you're going to engage in some significant philosophical and theological arguments. They're not answers that science can give us ... So, the idea that science could actually displace either philosophy or religion seems to me a complete nonsense.

Historical Perspectives

The full audio and transcript of the interview with Peter Harrison can be found on the ISCAST website at: http://ISCAST.org/interview/Harrison_interview_2017

… **Present Currents:**

Atheism, Secularism and the Spirits of the Age

4. Challenging the prevailing secularist narrative

By Scott Buchanan

The self-styled opponents of "myth and superstition," such as New Atheist Richard Dawkins, have been shrewdly peddling a few myths of their own.

In former times, secularism meant the state remained neutral in the face of competing worldviews and comprehensive claims about reality. Ideas could be freely trafficked in a pluralistic environment, whilst no one religion or creedal system could claim official establishment. Although people adhered to a minimum set of shared values—the better to preserve social and political harmony—all were permitted to enter the public square according to their own lights and their own convictions.

More recently, however, a new conception of secularism has arisen. Unlike its intellectual forebear, the contemporary model is neither neutral nor passive in regards to contrasting worldviews. Quite the contrary. In fact, it is largely built upon a fundamental antipathy towards what it sees as the unwarranted public influence of "belief." Scientists like Richard Dawkins and Neil de Grasse Tyson exemplify this view, whilst Australia is also home to its own tribe of new secularists. Proponents of this view devote themselves to a vision of the public square expunged of anything allegedly lacking in scientific objectivity. Much of this ire has been directed, of course, at religion.

The new secular project rests on two complementary claims: that certain value-systems—particularly those codified in religious traditions—are hobbled by a corrosive irrationality; and that secularists enjoy the benefit of an objective, unmediated view of reality. For the new secularists, there exists an irreconcilable division between these two realms—between a grounded, life-giving realism, and an enervating superstition. However, despite their increasingly widespread popularity, these assertions are quite unfounded.

Let's examine the first claim: namely, that religion is irrational. Dawkins encapsulates this view well when he condemns (religious) faith as "blind trust, in the absence of evidence, even in the teeth of evidence" (*The Selfish Gene*, 1989, p. 198). For him and those of his ilk, religion is bereft of rational and evidential justification. This isn't merely the claim that *this* religious adherent is irrational or *that* doctrinal formulation is without foundation;

it is, rather, the much stronger assertion that religion *as such* is rationally deficient—the product of delusion, wishful thinking or a stultified intellect. But this claim flattens out the diversity of religious belief and experience, in both nature and origin. It's impossibly broad, ignoring the rich intellectual traditions of some of the world's major religions, and the sophisticated arguments that have been developed to substantiate them.

For example, the medieval Catholic monk, Thomas Aquinas, was a skilled exponent of rational demonstrations of theism. In his "First Way" (a type of cosmological argument), Aquinas argues that the everyday objects of our experience, and their causal interactions with each other, furnish a base from which a person might reason, via metaphysical principles, to a sustaining cause of the very structures of reality. He saw that finite things possess latent properties that can only be "actualised" (that is, brought from the realm of the potential into the realm of the actual) by external forces; change within an object is the result of those forces acting upon it, whatever they may be. To take a simple example, a red rubber ball left in the sun will eventually turn a lighter shade of pink; place it near a hot flame, and it will, over time, change into a puddle of viscous goo.

According to Aquinas, these changes are part of larger, and more complex, chains of causation. Each member within that chain has only secondary causal power, simultaneously depending on earlier members for whatever potency it exercises. Delving down into ever-deeper layers of reality, the First Way takes one to its basic structures. Simultaneously, the First Way also argues against an infinite regress—that is, an endless ribbon of causal activity, stretching downwards *ad infinitum*. For Aquinas, it would be metaphysically "groundless," having nothing upon which to become extant. And, if so, then it must terminate in a fundamental cause, sustaining all else and actualising all secondary causes. Sitting at the foundational strata of reality means that this fundamental cause could not, in principle, be a part of it—as if it were merely some finite feature of our world. Rather, it would have to be the very source and ground of *all* being, the metaphysical basis upon which the world exists in the first place. And, for Aquinas, it would have to correspond to what people traditionally know as God.

Of course, new secularists might retort that most religious folk don't think this way, but rather construct their beliefs in a more unreflective manner. However true this rejoinder may be, it fails to realise that many arguments for, say, God's existence—no matter how intellectually demanding—actually build upon the *quotidian* experiences and intuitive impulses of ordinary people. Aquinas' own explorations depend on everyday observations in order to get off the ground. Other arguments of this kind are partly based on a person's ordinary (yet reasonable) reflections concerning causal principles, a sense of the transcendent, a belief in the world's rational intelligibility, and even its apparent contingency. As the theologian Keith Ward notes, belief in the kind of God Aquinas sought to substantiate plausibly fulfils many of these longings—"for God," he writes, "is ultimate reason ... [and] the only belief which gives reason a fundamental place in reality" (*Is Religion Irrational?*, 2011, p. 61). Such arguments may distil, challenge or stretch certain aspects of a layperson's unfocused understanding of theism. Nonetheless, they are not fundamentally at odds, and imply that the basic drives people possess towards the divine may be quite consistent with rational theistic accounts.

New secularists might still contend that such arguments simply fail to supply evidence for God's existence—and, therefore, lack any rational warrant for religious belief. For them, a reasonable belief is largely synonymous with what is empirically demonstrable. But as the philosopher Edward Feser has perceptively suggested, this criticism founders for the very reason that it adopts an *a priori* (i.e., non-empirical) assumption about what counts as "rational," "evidence," or "warranted belief." The scientific enterprise is merely *one* avenue towards knowledge and truth; other methods of rational inquiry exist, including mathematics and philosophy, which do not rely fundamentally on empirical observation. Moreover, the very assumptions scientific study takes for granted—the existence of the external world, its rational intelligibility, the reality of causation, or the general reliability of one's senses—suggest that such a project cannot even get off the ground without implicitly appealing beyond itself.

What, then, of the new secularists' other assertion: that they alone, as people free from the encumbrances of bias (both religious and otherwise), enjoy a transparent view of reality? How should one respond, say, when a Neil de Grasse Tyson argues we need a new "country"—*Rationalia*—whose constitution stipulates that public policy should be stripped of all value-statements, and formed on the basis of pure facticity?

Irony abounds, for De Grasse Tyson's own position represents as clear a value statement as one would want. To be liberated from all prior values and assumptions is intrinsically impossible, since no one makes enquiries about the world in a vacuum. As the missiologist and theologian, Lesslie Newbigin, has noted, human beings are inescapably bound by their finite vantage-points, and are conditioned by prior plausibility structures that reinforce or screen out certain patterns of thinking. Similarly, the sociologist and political theorist, Barrington Moore, Jr., wrote that,

> Human beings ... do not react to an "objective" situation ... There is always an intervening variable, a filter ... between people and an "objective" [event], made up of all sorts of wants, expectations, and other ideas ... (*Social Origins of Dictatorship and Democracy*, 1966, p. 485).

I've already noted that even those who prize empirical observation above all else must still begin with a received picture of the world. Secularists who tout the predominance of "facts," and who try and ground their views in an exclusive kind of empiricism, are unwittingly committed to their own set of plausibility structures—in this case, presupposing that reality can only be captured by the methods and processes of modern science. Again, the new secularist, just as much as the religious devotee, is inherently incapable of adopting a completely value-free position.

Additionally, facts by themselves can't do all that much; they need to be held together coherently, according to an overarching narrative or interpretive framework, if they are to mean anything beyond their own referents. The debate over abortion is a good example of this phenomenon. Modern science might be able to determine in great detail when a foetus begins to develop vital organs, when it is able to feel pain, and so forth. But how can it tell us if abortion is, under any circumstances, morally justified? How can it determine when, if ever, a baby with a developmental disability should be terminated? Even framing the questions in such terms is a category mistake: thanks to Hume's observation that one cannot derive an *ought* from an *is* (at least not without further argumentation), it's clear that simplistically trying to read prescriptive truths off descriptive data cannot be done.

Some, like Dawkins, think that one of the crucial questions regarding the morality of abortion is that of foetal suffering. Though important, such consequentialism is simply not the logical product of scientific enquiry. He proceeds to argue that the moment of birth forms a "natural Rubicon" between permissible and impermissible acts of killing. But again, how does the scientific enterprise lead to such a distinction? What essential difference is there between a child who has been in its mother's womb for eight months, and a child just born? Dawkins' line-drawing is arbitrary, having little to do with a pure, empirical appraisal of the situation. One might equally argue that conception marks the basic ontological transition from non-being to being, and is therefore the "natural Rubicon" one ought to use; indeed, everything subsequent to that epochal moment simply represents its unfolding. The point, however, is that these issues—the nature of personhood and the value one should ascribe to it—are fundamentally philosophical and metaphysical. Scientific enquiry *alone* cannot provide complete answers. Consequently, the secularist's much-vaunted neutrality dissipates, and she once again finds herself in the same boat as the religious adherent—compelled, that is, to rely on a basic array of presuppositions to guide her ethical analyses and prescriptions.

Conclusion

As much as the new generation of secularists would have us believe their claims regarding religion, truth and reality, it's clear those arguments are deeply unsound. It's therefore difficult to avoid the conclusion that attempts to squeeze religious and other value-laden convictions out of the public sphere do not proceed from innocent scientific or rational enquiry. Rather, those methods have been pressed into service to help prosecute an agenda possessing quite different origins. If this chapter has succeeded in anything, then it has at least shown that the self-styled opponents of myth and superstition have been shrewdly peddling a few myths of their own.

> An earlier version of this chapter first appeared at scottlbuchanan.wordpress.com/2016/12/09/challenging-the-secularist-narrative/. Reproduced and edited with permission.

5. Questions for Phillip Adams on his rejection of faith

By Stephen Ames and John Pilbrow

Stephen Ames and John Pilbrow respond to criticisms of belief in God made by writer, broadcaster and atheist Phillip Adams in an article in *The Australian* newspaper.

It is wonderful, mind boggling stuff. In 1998 Professor Brian Schmidt with his team from the Australian National University discovered that the expansion of the universe began to accelerate about six billion years ago—some eight billion years after the big-bang—due to "dark energy," and that the universe would go on expanding for ever into a "big freeze."

Astronomers had long wondered whether the universe would go on expanding into a big freeze or whether gravity would eventually stop the expansion and contract all matter into a fiery crunch, the "big fry." Schmidt shared the 2011 Nobel Prize in Physics for answering that question.

Writing in *The Australian* on 3 December 2011, writer, broadcaster and atheist Phillip Adams used this celebrated result to launch another tirade against God. Schmidt's work is not a problem for Adams who has long considered that we live in a meaningless universe—"without author or purpose." But he thinks this is bad news for believers in God and an afterlife: "Worshippers will be demanding their money back." Adams concludes by urging us to accept a number of "exhilarating" facts, which we will come to later.

The discovery of dark energy and dark matter is a grand puzzle to physicists. It portends a great revolution in our scientific understanding of the universe. We applaud them for their work and await a fuller explanation with great interest. However, this is hardly "bad news" for Christians. For us, God is both transcendent, beyond the universe, and immanent in everything—closer than breathing. Christian faith looks to the complete transformation of the universe by God—the coming of the reign of God in glory, when God will be "all in all." The prospect of the big freeze is not the last word about the universe. In fact, Christians believe God, not natural science, has the last word and indeed the first word about the created universe. Yet we believe faith and reason both come from God and so we should expect them to be consonant, not in conflict when both are properly understood.

For example, we would like to understand how this dark energy and the big freeze are an intelligible part of the divine economy for the whole creation from beginning to end.

Here is another example. Adams often says, as in his *Australian* article, "If God created everything, who or what created God?" He claims to have found no satisfactory answer to this question which he came to as a little boy. We judge it to be a mistaken question. If God created everything, then God is the cause of all causes. So any proposed contender for being the "creator of God," has already been created by God. Adams doesn't understand the concept of God that he is using. There are powerful reasons why people don't believe in God—e.g., the problem of natural evil—but Adams' boyhood question is not one of them. Phillip, perhaps it is time to move on.

Adams' exhilarating facts

Let's look at Adams' four exhilarating "facts."

(1) We are not the centre of the universe. Christian theology has no problem with that claim, either scientifically or theologically. Scientifically it is to be recognised that this insight comes out of astronomy/cosmology since the time of Copernicus and Galileo. This was a truth that the world and the church needed to understand. Theologically, it is God who is the centre and the circumference of all that is, not human beings. According to Psalms 96 and 98, all the trees, mountains and seas will exult with joy at the coming of God. The whole of creation is seen and experienced differently when understood to be centred in the God revealed in Christ.

(2) Our human lives are brief and inconsequential. Brief, yes, but inconsequential? Notice how Adams pronounces a "last word" on human life. Phillip, what is it that makes you think we are inconsequential? Is it the big freeze ending? But this is rather like treating the debris from a wonderfully wild party as the meaning of the party! What might change your mind? Paul Davies is one among many scientists who see human consciousness as the place where the universe becomes self-aware. Is that inconsequential? Hardly, given the amazing power of human consciousness. So any story about the universe has to explain consciousness. But what kind of story can do that? It depends on what we say about consciousness. Might it include things that go beyond any scientific story? Think of the consequences!

(3) We are free to live our lives without risk of damnation or some religious equivalent of Frequent Flyer Points. Phillip, given what you believe about God this makes perfect sense. However, the gospel says God so loved the world that he sent his Son so that all who believe in him should not perish but have eternal life.

(4) The only meaning our existence has—or the existence of the entire cosmos—is the meanings we choose to give it. Might I suggest love? And the joys of curiosity? Christians will agree with something like the first suggestion, in part because of the good news of God's love for the whole cosmos revealed in Christ. But Phillip, tell us how you can say human life is inconsequential yet suggest love is capable of giving meaning to the existence of the whole *cosmos*? You make different claims in so-called fact 2 and fact 4. Both cannot be true.

Some Christians will agree with the second part of fact 4 because even without Christ they may well have discovered much about the value and the truth of love—like the Good Samaritan, and those "sheep" on the right hand of the king in Matthew 25:31–46, who clearly had never heard the gospel but acted to feed the hungry and clothe the naked without any thought of God, Christ or their own salvation. But we are speaking about Christians who agree with Phillip! They must have found a way to see the existence of the whole universe as having meaning in the light of human love. How do they do that? How does Phillip do that? What is it about the ordinary human experience of love, (even allowing that it can be caught up with lust), that gives it the power to help us see the whole universe as meaningful and for us to choose to believe that?

At the 2010 Global Atheism Conference, held in Melbourne, Australia, the ABC's Robyn Williams cunningly introduced an old issue by joking about being tempted, as an atheist, to consider coincidences as perhaps "signs" of God's activity, as religious people often do. But then he abruptly changed focus by quoting from *The Rape of the Congo*, about an incident of appalling slaughter and rape. The dreadfully abused woman watched her dearly beloved husband as he was slowly and painfully killed. With vehemence Williams asked where God was in this and countless other occasions of human violence. His answer—God didn't intervene because he doesn't exist!

A Christian response would point to the crucifixion not only of Jesus, but of countless others who were horrifically executed by Rome in defending its empire. There were no interventions, not even one to stop the killing of a man who according to the gospel story was the innocent, incarnate Son of God. It is in fact God who ends up a victim of human violence.

But what does Williams' story show us about the truth and value of love? The revulsion at the violence shows overwhelmingly that the unconditional worth and dignity of the man and the woman and their relationship had been utterly violated. If you take this account of revulsion as a clue to reality, then you cannot believe that everything is conditioned by everything else. For then nothing would be unconditional and your revulsion would merely be another piece of conditioning. Is that all it is? Whereas, if you take this as a clue to reality, your worldview must include something transcendent, beyond all the conditioning, something that you are intimately in touch with in that revulsion.

Likewise, we think love includes the recognition and response to the unconditional worth and dignity of another, whether a stranger who crosses your path or the person with whom you share your life. If you also take this as a clue to reality then again your worldview must include something transcendent, beyond all conditioning: something that you are in touch with in that love, something which itself is of unconditional worth.

We do take this as a clue to reality which tells us something about the kind of cosmos which has produced us. It is a clue to ourselves, and to the transcendent God immanent in our lives, often unrecognised. This is one account of human love as pointing to the meaning of the existence of the universe.

Phillip, we would like to hear your account. You suggested another source for us attributing meaning to our existence and the existence of the whole cosmos; it is the joy of curiosity. Curiosity operates on many levels. It includes human inquiry in all its forms. One very powerful form is in the natural sciences, which for all their technological applications, are still motivated by curiosity as a pure desire to know. What would we find if we inquired into human inquiry? What is it about human inquiry that could possibly serve to suggest that it makes our existence and the existence of the universe meaningful? We would love to discuss this with you.

6. Lawrence Krauss takes atheism to a new low

By Chris Mulherin

In a debate that took place in Melbourne, Australia, in 2013, Lawrence Krauss, astrophysicist and polemicist for a rationality that includes nothing but science, clashed with William Lane Craig, a philosopher who promotes Christian faith as the most reasonable worldview on offer. Chris Mulherin attended the debate and interviewed Krauss.

In the cultural battle between New Atheism and faith, how does one act justly, love mercy and walk humbly with God when confronted with the arrogance of people who have no time for charity, humility or other traditional virtues?

Such was my spiritual turmoil when two titans of the global "God wars" crossed swords in Melbourne in 2013.

Craig and Krauss are like chalk and cheese in more than their views on the God question. While Craig was respectful and focussed on clarifying the argument, Krauss played the entertainer with little time for serious discussion. He was disparaging and aggressive as he paced the stage and played to the gallery, convinced that there was no substantial argument to be had and secure in his sure knowledge that science had done away with the God hypothesis. When it was his turn to listen, he seemed uncomfortable as he squirmed in his seat with pained impatience or grimaced with defiantly-crossed arms or shook his head in disbelief at what Craig was saying. In a pre-debate interview Krauss tempted me to abandon Christian charity as he compared Jesus to Hitler—see more below—and pompously proclaimed the God question was irrelevant.

Arguments about nothing

Krauss is a recent arrival on the atheist public speaking circuit, having sprung to fame with his views on "nothing." His 2012 book, *A Universe from Nothing*, attempts to answer the old conundrum of its subtitle: *Why There Is Something Rather than Nothing*. His answer is that physicists can now explain how the universe was brought into existence from nothing plus the laws of nature.

Krauss has riled atheists and theists alike with his ludicrous claims about "nothing." The problem is not necessarily his science; it is that Krauss redefines the word "nothing" so as to include the laws of physics. As the philosophers have been quick to point out, the laws of physics are "something" and not "nothing"; Krauss is playing a word game, which, despite his pretensions, has no implications for whether there is a creator.

Here's an example of his doublespeak: "Nothing is a physical concept because it's the absence of something, and something is a physical concept." Now, it doesn't take a philosopher to point out that there are lots of "somethings" in the world that are not physical. My love for my wife, the number 42 and God himself are all something rather than nothing, but none of them are physical concepts. Surely Professor Krauss can't be so green about philosophy and theology as he seems to be? My conversation with him proved otherwise.

Kraussian faith

Krauss's view on faith is the standard New Atheist line that equates all religions to belief in the tooth fairy or Bertrand Russell's "celestial teapot." He believes that "religion is dying on its own" and thinks that his contribution to the cause is to "encourage people to replace the kind of things they get from religion with things that are related to the real world and not myths and fairy tales." I asked him what he saw as the role, if any, of Christianity in an increasingly secularised culture. His response was dismissive and pragmatic:

> Well, it gets in the way. I think the role is to provide some sense of community and support for people. The point is, do you need religion to do that? Right now the best role Christianity can play is to support systems that try to bring people together, and then get out of the way.

Krauss is unequivocal: there is a fundamental conflict between science and faith. Religion for him is baloney and the only way he can see the world is through the lens of science:

> Science is incompatible with the world's major religions; all of those are, from a scientific perspective, nonsense. God is irrelevant. People seem to think it's an important question; it's not an important question to scientists. God isn't necessary to discuss the universe.

Science as morality

Assuming that Krauss's "science is everything" view held water, I asked him what resources his atheist naturalism could draw on to provide meaning and moral guidelines. Again his answer revealed no reflection on the demarcation between science and other realms of human thinking. For him, not only does science provide understanding of the physical world but it also offers its own spirituality and is the foundation of morality:

> Scientific empiricism—rational thought combined with empirical enquiry, which is the way we learn about the world—brings much more meaning and spiritual wonder than religion ever does. The awe and wonder of the real universe is every bit as spiritual, in fact more spiritual, than the bland and boring and wrong stories in the Bible. But it has the great benefit of also being true. Science can enhance your appreciation of your place in the cosmos and of course provide a much sounder moral framework—a framework for determining what's right and wrong—than religion.

This unwitting logical flaw is what philosophers call the naturalistic fallacy; it's the attempt to draw moral conclusions from the facts of nature. It's what lies behind so called "scientific" programs of genocide and eugenics and it's what you get when science is allowed to rule in every area. Science as a worthy vocation for studying nature becomes *scientism*, a worldview which proclaims that all that is not empirical is nonsensical.

When I asked how science provides moral foundations, Krauss fell into philosophical gibberish: "There's no doubt if you look at morality in the modern world it's based on science; it's not based on religion." While he tritely said, "you can't make a decision about what to do unless you know the implications of your actions," which science can tell us, he then suggested that science was responsible for ending slavery, the emancipation of women and the end of homophobia.

Turning science into a worldview

I wondered if anything could possibly unsettle Krauss's supreme confidence in only the things that science could comment on. Again his answer revealed his pragmatism as well as an impoverished understanding of what might count as evidence.

> My convictions are, if there is empirical evidence for something then I'm willing to accept that fact. Nothing is going to unsettle that

because that works. If I looked up tonight and the stars rearrange themselves to say "I am here" then it would be worth thinking about. But as there has been no single shred of empirical evidence for any deity or any divine intervention in the history of the universe, there's no reason to worry about it.

Such flippant ignorance casts all religion, all history, all moral discussion into the flames. After all, there is no "empirical evidence" of the Kraussian sort for the existence of any historical figure, the equality of human beings, the wrongness of torture or the meaning of life. In the words of Richard Dawkins,

> In a universe of blind physical forces and genetic replication, some people are going to get hurt, other people are going to get lucky, and you won't find any rhyme or reason in it, nor any justice. The universe we observe has precisely the properties we should expect if there is, at bottom, no design, no purpose, no evil and no good, nothing but blind, pitiless indifference.

Delusions of belief

When asked about those respected scientists who are also serious believers Krauss resorted to the extraordinary ruse of suggesting that scientists who are religious are mentally deficient. "You can be a scientist and a Christian, but to do that you have to suspend your disbelief," he said. Again he echoes Dawkins' view that "religion poisons your ability to use your brain." (When asked about the faith of world-ranking palaeontologist Simon Conway Morris, Dawkins' reply was, "When Simon speaks of religion he leaves his brain behind.")

Krauss is paradoxically pessimistic about human rationality, invoking the White Queen's advice to Alice, although not realising that in doing so he undermines his own thinking, including his convictions on the God question:

> People can believe diametrically opposed things at the same time. We're hard-wired to be able to do it; we all do that. I like to say we all believe ten impossible things before breakfast. I think that scientists who have a strong belief just put it aside when they are doing their science. They say "I'm going to be guided by what the science tells me about the natural world but I won't let it infringe on this belief that I have about an imaginary guy in the sky."

This convoluted thinking is why Krauss and many New Atheist showmen are derided by serious thinkers, atheists included. When challenged about his shallowness, he obliged me with a memorable quotation; I asked if there was a danger for a physicist wading into the depths of philosophy and theology.

> No, because there are no depths of philosophy and theology. They are very shallow. I don't wade into philosophy and theology; I don't need to. You don't need to know anything about philosophy or theology to do physics. Or to understand the universe. I don't wade into the alien abduction literature either because I don't need to.

At this point I was tempted to put off the mantle of politeness. But first, a final question: "What do you make of Jesus?" I asked.

> I see him as an important historical figure in the same sense that Muhammad was. But there were other figures that were historically important—like Adolf Hitler. I don't view him any more profoundly than anyone else.

Such is the impoverished worldview and confused thinking of those who lead the New Atheist assault on the gates of heaven.

> The full audio and transcripts of Chris Mulherin's interviews with Lawrence Krauss and William Lane Craig are at: www.skandalon.net/KraussCraig. You can watch their debate at: http://youtu.be/7xcgjtps5ks.

7. Thank God for "The Rise of Atheism" convention

By Chris Mulherin

"The Rise of Atheism" global atheist convention was held in Melbourne, Australia, in March 2010. What can the church learn from the conference?

What could 2500 atheists and I have in common? More than expected, I realised, when I joined the "maddened" crowd at the Rise of Atheism convention. Yes, *en masse*, the atheists were mad at religion and mad at religious meddling in society and politics. They were especially mad at Christianity: mostly its historical atrocities and its sexual norms and abuses.

The prime attraction was Richard Dawkins, author of *The God Delusion*, and one of the "four horsemen" of the so-called New Atheism. I was keen to engage with the issues and to judge for myself the mood of this movement. At times I was surprised, at other times disappointed, at other times challenged.

I was surprised by how central religion was for the atheists represented. It's remarkable that, try as the New Atheists do to distance themselves from religion, there is a sense in which they are defined by it, by their anti-religious and anti-theist stance. I heard echoes of Alister McGrath's suggestion that "Western atheism now finds itself in something of a twilight zone." McGrath comments, "Once a worldview with a positive view of reality, it seems to have become a permanent pressure group, its defensive agenda dominated by concerns about limiting the growing political influence of religion."

I was also surprised that, despite our apparently opposed belief systems, I shared common ground with these committed "true non-believers." Like Christians, the atheists have taken their stand. In a bizarre sort of way, it was good to be amongst people who have no truck with relativism or with a postmodernism that turns truth into plasticine.

At times, the claim was made that atheists are only united around their non-belief in God. But most delegates shared more in common, especially the conviction that science offered the way, the truth and the answers to life. They also shared a frustration at our not-secular-enough society. While Christians feel marginalised at times by secularism, it was interesting to realise atheists' angst at encountering religion at every turn of politics

(Christian politicians), education (chaplains in schools) and law (tax exemptions for religious organisations).

Philip Adams, perhaps Australia's most vocal atheist, surprised me with his call to moderation and to serious conversation with those who do not share atheists' non-faith. Rather than focussing on emptying the churches, atheists should unite with people of similar political and global concerns. It was a call to humility and balance, albeit with the constant undertone that while Christians and others might be committed to the right causes, their beliefs are nevertheless foolish. "People of religious faith are more to be pitied than blamed," he said.

Adams warned about stereotypes and prejudice: "We do not have a monopoly on decency." He also warned those who refuse to accept "the possibility that Christians, for example, can be significant social reformers," recognising that, "it was Christians following Wilberforce that worked mightily to destroy the slave trade." Just because Christians lived with the hope of "frequent flyer points at the end" is not a reason to doubt their ethical integrity, he said.

While Adams and others made for thoughtful listening, I was disappointed overall with the program. I hoped for discussion about the nature of science, but this meeting was not about debating issues of faith and non-faith. There was little attempt to argue for atheism as a (non) belief, nor to defend the view that science is the *only* source of truth. The message was that religious people are simply misguided and unwilling to accept clear empirical evidence, however, it was a simplistic "science" that dominated proceedings. The program included comedians and, despite suggestions that Christians have hang-ups about sex, some were fixated on the subject.

Richard Dawkins gave a fascinating talk about evolution and the finetuning of the universe. While the science was riveting, his characteristic disdain for believers was also on show. An example: while he respects Cambridge palaeontologist Simon Conway Morris as a scientist, Dawkins says they part ways because Conway Morris sees convergent evolution "as evidence for his weird belief in Christianity." Asked how scientists could be Christians, Dawkins replied: "Religion poisons your ability to use your brain."

Yes, for the New Atheists, mutual understanding is impossible and religion is a virus to be eradicated. For them, theology is the time-wasting study of nothing. Does this incompatibility extend to living together in the world? Unfortunately, for some it does. For various speakers, there seems no possibility of a sane and harmonious future until the world is rid of religion.

However, all atheists cannot be tarred with the same brush. The anti-theist New Atheism is not representative; many atheists and agnostics do not share either the zealous non-faith or the political agenda of this "new" variety of an old tradition. Many would not be happy with Dawkins' description of brain-poisoned believers. Atheist philosopher Jim Stone recognises that believers are often very good philosophers respected by their atheist colleagues. He says: "The people I don't like are the New Atheists because they don't seem to realise that the people with whom I must contend even exist."

Philosopher and animal rights activist Peter Singer spoke on "Ethics without religion," examining the roots of morality. His answer was equivocal: on the one hand, morality is perhaps hard-wired in the brain. But Singer warned: "We shouldn't fall into a trap of thinking that a natural response is necessarily right." Our moral judgements have evolved for situations we are familiar with, and might give us the wrong answers in new situations. For example, racism might be inbuilt, but it "is something we need to get over because the world is a very different place to what it was." In the end, Singer's measure for morality is rooted in a version of the "pleasure principle," which maximises the well-being of all sentient beings.

What about the challenges for Christians? What can we learn?

First, it is salutary for the church to listen to its critics. Although not verbatim, the convention accusation was, "you have not acted like Christ." I see two challenges: one is that of respecting the opinions of those who do not share our faith; the other is dealing with sin in our camp with a trust-building transparency. Positive comments from atheists were posted on my ABC blogs and I enjoyed frank conversations, including an amicable one with well-known Melbourne atheist Catherine Deveny.

I wondered how much of the anger at the Convention was rooted in pain for crimes and misdemeanours committed by the church and I remembered that Jesus rejected no one except those who thought themselves righteous.

Secondly, Christians ought to thank God for the dialogue made possible by atheism. Conversation with atheists throws the spotlight on the distinctive features of our faith. Beliefs and agendas are on the table for all to see. This is a challenge to our temptation to hide the light under a bushel for fear of causing offence. Christianity is *not* simply one option amongst other compatible belief systems. Engaging with atheists reveals that truth.

Thirdly, while Christianity's unique claims are highlighted by the dialogue, that is no reason to participate in culture wars, premised on the idea that one group must dominate. We need to examine the implications of living in an increasingly secular society where a harmonious future will only be forged through mutual tolerance. Trust can be built only when beliefs and values are clear and when all parties accept the limitations imposed by an impartial democratic state. As Christians, we ought to preach the gospel in word and deed; we ought to persuade others, offering good reasons for our hope. But coercion and manipulation are ungodly and bring disrepute.

Thank God for "The Rise of Atheism" and for conversations with atheists who provoke us to reflection.

8. The New Atheists and fundamentalists need to move on

By Alister McGrath

The old narrative of the conflict of science and religion is ideologically driven. It's time to recognise that human beings need both science and religion if they are to flourish, and move on from the simplistic and ignorant approach of some New Atheists and fundamentalist Christians.

My first love was science. I've never stopped loving it.

Science is great at taking things to pieces to find out how they work, helping us grasp the mechanisms underlying the processes we see around us in the world. We long to be able to make sense of what we see and experience in the world, seeking deeper patterns lying beneath what we experience and observe.

To my delight, the beauty of the night sky could be supplemented by something even more beautiful—the mathematical representations of the cosmic processes that brought it into being in the first place, and sustained it thereafter. And that, I then believed, was all that could be said, or needed to be said.

I have to confess I then had a marked cultural distaste for religion, sharpened by what I concede was intellectual arrogance. Religion was for intellectual losers. As a sixteen year old lover of science, I believed that atheism was my only legitimate option. Science proved its beliefs; religious people just asserted things which had no evidential basis, and seemed to be designed to console them with spurious delusions of meaning.

So what if science couldn't provide answers to what Karl Popper called "ultimate questions"—such as: Why are we here? What's the point of life? Science didn't have the answers to these questions precisely because there were no valid answers to be had. People just made these things up, whereas I was prepared to stare a meaningless universe in the face, and confront its existential bleakness.

So I focussed on studying the natural sciences and mathematics at my school in Belfast, Northern Ireland. I had set my heart and mind on a goal, and made it my sole object to achieve it. I loved chemistry, and Oxford University had the best chemistry course in Great Britain. I had to stay on at school for an extra term to take Oxford's entrance exams, but it was worth it.

Present Currents: Atheism, Secularism and the Spirits of the Age

In December 1970, I learned that I had been awarded a scholarship to study chemistry at Oxford from October 1971.

So what would I do while I waited to go to Oxford? I decided to stay on at school for a further two terms. It would allow me to develop my specialist knowledge of chemistry, and learn Russian and German—at that time, the two major international languages of chemistry, in addition to English. I did not realise it in doing so, but those two terms would raise fundamental questions about my take on science that would change the course of my life.

I don't remember exactly when I began to read books about the history and philosophy of science. The school library had a small collection of works in this field, and it didn't take me long to devour them. As I read them, however, what I had taken to be a simple and straightforward account of science and its methods began to crumble. I realised that I had only a schoolboy's grasp and understanding of science. The real thing was much more complicated and unsettling. I slowly came to realise that I had fallen in love with a hopelessly simplified account of science. I now had to learn a lover's hardest lesson—that there is a lot more to the beloved than what initially meets the eye, and that true love is only deepened by knowing the beloved *properly*.

Looking back, I can still recall the key insights I then acquired, and appreciate their impact on my life and thought. They didn't lead me to Christianity; they just destroyed a simplistic and naive account of science which prevented me from taking Christianity seriously.

For a start, I realised that I had failed to appreciate that science is a research method, not a fixed body of ideas. Science is on a journey, and changes its mind over time, partly through acquiring new data, and partly through continued theoretical reflection. The astronomer Carl Sagan (1934–1996) made this point elegantly:

> Science is much more than a body of knowledge. It is a way of thinking ... It counsels us to carry alternative hypotheses in our heads and see which ones best match the facts.

That means that the theoretical deliverances of the natural sciences are provisional. It's an interim report on where we are at the present, knowing that we may be somewhere else in a generation's time. Back in 1916, the scientific consensus was that the universe had always existed; now, we speak of it having a beginning.

This is philosophically perplexing. A scientist has to commit to certain theories, knowing that at least some of them will be shown to be wrong in the future. That's why Michael Polanyi's *Personal Knowledge* (1958) is so important to reflective scientists. Polanyi (1891–1976) was a Hungarian chemist turned philosopher, who found himself to be increasingly troubled by his need to commit himself to what he believed (scientifically) to be true, while knowing that some of this would later be shown to be false. He argued for the need to speak of science as "personal knowledge"—not absolutely certain, yet still capable of eliciting justified belief.

Scientific knowledge thus involves our personal—and fallible—judgement that certain beliefs are reliable, and to be trusted. Polanyi insisted that we must understand that commitment to beliefs—scientific or otherwise—inevitably transcends the evidence underlying them.

Second, I began to realise that the historical evidence for the popular stereotype of the "warfare" of science and religion was really rather inadequate. The scholarly destruction of this stereotype, of course, lay in the future. The massive historical revisionism which forced such a radical rethinking of this "conflict" or "warfare" model really began to emerge in the 1990s. Yet hints of it were already there, back in the late 1960s. As leading scholars in the field—such as Australia's Peter Harrison—have shown, there is no fixed relationship *of any kind* between science and religion, and most certainly not that they are, or need be, in essential conflict. Their interaction is rich, interesting and *complicated*.

Sadly, the movement we know as the "New Atheism" largely ignores these two points, tending to dismiss them as trivial concerns raised by people who feel threatened by science. Happily, this movement is now receding into the past, allowing us to get back to a constructive and above all *informed* discussion about the relation of science and faith. Sure, it raised some good questions. But the answers it gave were hopelessly simplistic. They may have seemed plausible in a simpler past. But not now. We need to move on.

Mapping human identity

In my book *The Big Question*, I explore my own transition from atheism to Christianity, and the conceptual frameworks for relating faith and science that I developed over a period of 40 years. This book is an interim report from the frontiers of science and faith, a promissory note which can never aspire to be "finished" or "perfect," partly because the fields are moving, and partly because there is too much for one person to take in and assimilate.

Present Currents: Atheism, Secularism and the Spirits of the Age

At its heart, the book is a plea for a civilised dialogue between different positions, opening the door to an enriched vision of reality. There are some diehards within both the religious and scientific communities who would resist this move, fearing intellectual contamination or loss of focus, or who persist in believing that science and faith are locked in some kind of eternal warfare. Yet we have moved beyond these outdated isolationist ideologies, which rest on highly questionable foundations.

In the book, I try to map out what form these discussions might now take, in the light of the shifting of the historical and philosophical tectonic plates that resulted from the massive research literature which has transformed the field in recent decades.

In one sense, the approach I set out in *The Big Question* is not new at all. It's a retrieval of an approach developed in the Renaissance, one of the most creative periods in European intellectual and cultural history, reworked to take account of the changing philosophical and scientific landscape of the twenty-first century. It's time to get this approach back on our cultural agenda.

I propose a principled and critical interweaving of narratives to help us deal with the "ultimate questions" that persistently refuse to go away in our cultural discussions. To answer these properly, we need to bring together multiple approaches, and recognise the existence of multiple levels of meaning—such as purpose in life, values, a sense of individual efficacy, and a basis for self-worth.

The approach I adopt is basically about recognising multiple perspectives on a complex reality, and multiple levels of explanation. It's an elaboration of the kind of approach proposed by the philosopher Mary Midgley, who insisted that the complexity of reality required multiple research methods and approaches. We need "many maps, many windows" if we are to represent the complexity of reality. "For most important questions in human life, a number of different conceptual tool-boxes always have to be used together."

Sadly, the New Atheism demanded that its limited and limiting approach to reality be seen as exclusively right. It was an unwise move. As the atheist philosopher and biologist Massimo Pigliucci pointed out in a recent study, this commitment to a narrow scientism was bad news for both science and atheism. He urged his readers to break free from the stranglehold of this deficient form of atheism, and adopt something more sensible and realistic.

My view, set out in *The Big Question*, is that science and faith can provide us with different, yet ultimately complementary, maps of human identity. And we need both if we are to flourish as human beings and lead meaningful and fulfilled lives. That doesn't make our need for meaning right—but it does make it *human*. Both science and faith are prone to exaggerate their capabilities. Religion cannot tell us the distance to the nearest star, just as science cannot tell us the meaning of life. But each is part of a bigger picture, and we impoverish our vision of life and the quality of our lives as human beings if we exclude either—or both.

Rethinking "humanism"

That's why I challenge those who use the term "humanism" to refer to an anti-religious way of thinking and living. I have no problem if they call this "secular humanism." But "humanism" is about what gives us identity and meaning as human beings. If the cognitive science of religion is right, and it is *natural* to be religious, how can anyone use the term "humanism" to designate a necessarily anti-religious or secular perspective?

It's time to suggest that "secular humanism" needs to name itself for what it really is. Any form of humanism ultimately rests on an understanding of what human nature really is, including what longings, desires and aspirations are naturally human. A Christian humanist declares that humanity finds its true goal in discovering God. A secular humanist declares that humanity finds its true goal in rejecting God. But to pretend that "humanism" is *necessarily* "secular humanism" is indefensible.

The word "humanist" had no such overtones or associations at the time of the Renaissance; they emerged in the twentieth century. It's surely time to move on. Christian humanism is alive and well, even if secular humanism pretends it doesn't—and can't—exist. I expand this point in *The Big Question*, and will be returning to it repeatedly in the future. There is a real issue here about the pretentions to exclusivity on the part of some—but, I must stress, not all—secular humanists.

Beyond ridicule, to respect

There's much more that needs to be said about all these issues, and I open up these and many other issues in *The Big Question*. The dialogue between faith and atheism is potentially one of the most important and most constructive for the future of Western culture, as we debate issues of meaning, morality and social cohesion. I long to see a respectful dialogue, for example, between secular and Christian visions of human identity and flourishing—in other words, between secular humanism and Christian humanism.

There is, of course, a big obstacle to any such serious dialogue at the moment. Fundamentalists on both sides dismiss any suggestion of dialogue or respect as a mark of treason, contamination and collusion. I think, for example, of the New Atheist internet trolls who seem to have fallen victim to one of the most distasteful features of Christopher Hitchens' polemic—namely, the deployment of ridicule and vilification in place of reasoned and evidenced argument, sensitive to the complexity of things. It reminds me of Plato's criticism of the Athenian politics of his day, in which "rudeness is taken as a mark of sophistication."

I would never dismiss atheism as ridiculous, or atheists as fools. I used to be an atheist myself—and that leads me to respect atheism, not to treat it with contempt. Sure, there are some very strange atheists, mirroring the weirdness of some religious people. But you can't treat mavericks on the fringe as if they're the mainstream. We need a principled, respectful dialogue which actively seeks out the best representatives of a position, and engages them constructively and critically.

Sadly, the most distinctive feature of the New Atheism is its propensity to parody and vilify. It got some headlines back in 2006–2007, when it had some novelty value. But happily that's worn off. Now that the New Atheism is on the wane, maybe Christians and atheists can get back to the kind of serious conversations that are so badly needed.

The Big Question tries to convey the sense of delight and intellectual fulfilment that I find in exploring the vibrant vision of reality that results when science and faith are allowed to critique and enrich each other. The old narrative of the conflict of science and religion is now seen as historically underdetermined and ideologically driven. Its spell has been broken. It's time to move on, and find and embrace a better approach, such as the narrative of enrichment I propose for discussion. Maybe it's wrong. But that just makes it all the more important to have critical and constructive atheist voices in the conversation.

> Alister McGrath's most recent book is *The Big Question: Why We Can't Stop Talking about God, Science, and Faith*, published in the United Kingdom as *Inventing the Universe*. This chapter first appeared as an article on the ABC Religion and Ethics website. See www.abc.net.au/religion/

9. The "spirit of the age" is out of touch with God's call

By Stephen Ames

Much in modern life—whether digital technology and communication, the market economy or the pace of life—creates a dissonance with belief in God and how Christians are called to live.

The church shares in God's mission in and for the world. This mission includes giving an account of the kind of world in which we live. Not just in the terms of the various accounts in circulation but also in terms of the "divine economy" for the whole universe from creation to consummation.

What kind of world are we living in? Near where I live Telstra has a huge billboard facing Nicholson Street, Carlton. It reads, "Let us explain the NBN in the old fashioned way—face-to-face." What is the latest fashion? Digital communication to the absent other. Of course even Telecom (prior to Telstra) promoted this with its vision of a phone in every home. But the difference now is that digital communication is becoming the dominant form of communication and it renders the face-to-face way old fashioned.

The billboard shows us a culture saturated by the natural sciences, the ever-new digital technologies and the global market. From these many people distil some fundamental themes which have become widely taken for granted about the kind of world in which we live.

Firstly, from the natural sciences many people have formed the view that the universe has no objective purpose or value.

Secondly, digital technology functions as the new *lingua franca*—the common language of our day and a "first philosophy"—the way we are to think and talk about what is real, about how things work, about what is possible and the power to make it happen.

Thirdly, it is widely taken for granted that the personal arises wholly from impersonal natural processes (biological and neurological) and is shaped in and by society.

Fourthly, "value" or "worth" is predominantly "instrumental value." Something has worth or value as a means to a desired end, until the next latest object of restless desire comes to light. This form of worth becomes our daily wor(th)ship represented by the market. The ideas of unconditional value, intrinsic value or irreplaceable value fade.

These four themes indicate something of the human world we have produced from the good gifts of God the creator of the universe and of humankind as the image of God on earth. There are many resonances and dissonances between the life to which God calls human beings and the kind of life we have produced.

The mission of the church includes acclaiming the resonances and redeeming the dissonances. What does that mean for the four themes just identified?

The place of the natural sciences

The natural sciences are indeed one of the resonances between the gift of the universe created freely and rationally by God and human responses. This is shown both by the historical origins of the natural sciences first under Islam and then afresh in Christian Europe. It is also shown in the coherence between various key themes in Christian theology and the doing of science.

There is a dissonance, not with the natural sciences, but with turning them into a philosophy—scientific naturalism (scientism), which says: all there is, is what the natural sciences say there is. Since the natural sciences rightly speak about the natural universe, not about God, this *philosophy* entails atheism. This is a deep dissonance. The widespread confusion between this *philosophy* and the natural sciences *as sciences* is another dissonance. A redeeming move would be a review of educational practice in secondary schools that asked to what extent this confusion is hidden or brought to light, and even criticised, because there are substantial philosophical criticisms.

Two more dissonances. With very few exceptions scientists do not see evidence for any purpose in the natural processes of our universe. Neither "Fine Tuning" nor "Intelligent Design" are convincing. This appears to be a powerful dissonance against the belief that God creates this universe for a purpose—likewise the strongly felt dissonance between natural evil (tsunamis, genetic disorders, for example) and the gospel that God is love. Are there possibilities for redemption? It is possible to "back-engineer" technology with its blind natural processes to identify the purpose for which it has been constructed. A similar possibility is available using the work of atheist physicist, Victor Stenger on the laws of physics. Surprisingly, this "back engineering" concludes the universe is structured according to these laws in order to be knowable by empirical inquiry. Redeeming the dissonance of "natural evil" begins by answering the question, "Why would God create a universe *ex nihilo*, sustain it in existence, and within the universe still use evolution?"

Digital technology and the human person

The deployment of digital technology in a way that helps render face-to-face communication "old fashioned" is a deep dissonance for humankind created in the image of the triune God. It can be redeemed by the practice of communities attending to personal, co-present communications as essential to our humanity; and Christians working to promote faith communities as a resonance with the reality of the triune God and the flourishing of humankind across the planet.

Digital technology is used to re-write the construal of what is really real. Here it merges with the scientific naturalism mentioned earlier. It is seductive because of the power it gives us to open many possibilities to transform our relationship with nature, and to engineer genetic transformations taking us beyond anything recognisably human.

The next two themes concern the human person and the market. Here I can only note that at some point they intersect. Economists and marketers all make assumptions about human beings in a properly functioning market. The economy in which we live and move and have our being is a social movement ordering itself to better make people fit its assumptions. The economy has to endlessly go on growing. We all have to be consumers who consume ever more goods and services, without ever having enough. A condition for us "wearing" this construction of the person is that we have spiritual needs that are turbo-charged by the conditions of contemporary life: I refer to the diminishment of satisfying face-to-face communication, the pace of life, the fading of unconditional value noted above, and the loss of transcendence in daily life. These spiritual needs are misinterpreted as material needs which can be met by the latest goods and services provided by the market. In fact they cannot. This produces consumers who can never have enough, without really knowing why. This amounts to an exacerbated restlessness of the human heart that cannot find its rest in God, as Augustine taught long ago.

Two deep and connected dissonances, with many consequences, are the increasing gap between rich and poor and the reality of climate change. Many people point to the resonance that many millions of people have been lifted out of poverty. That is indeed a good, but it points to how many more might be lifted out of poverty by a no less productive economy whose benefits are differently distributed, or indeed by a different economy. For example the inclusive growth promoted by the Brotherhood of Saint Laurence or possibly Tim Jackson's *Prosperity Without Growth* (Jackson is Professor of Sustainable Development at the University of Surrey, UK). A redemptive move would be to make reducing the danger of climate change a central concern and to ensure the equitable production and distribution of food for the world. In both, Australia could take the lead, supported by the church's prayer for daily bread and concerted action.

A final concern is the sense of lived time as accelerating. The "touch and go" facility for financial transactions is just part of the "accelerating" lifestyle, in which the metaphor of "acceleration" refers to our experience of cramming more and more events into every day and being subject to turbulent change across our society. Inevitably this means we must attend more to the "surface" of life and relationships; in our 24/7 world there is no time to go deeper. The "spirit of the age" becomes the taken-for-granted spirituality informing our lives, whether we are religious or secular. This is a dissonance that intersects the other dissonances. Its redemption calls for a diagnosis of what drives the accelerating pace of life and will also surely draw on the riches of Christian spirituality translated into a liveable practice of being still and knowing God. A place to start might be with a simple "rule of life," an intentional pattern of discipline by which we aspire to live.

10. Alain de Botton's *Religion for Atheists* and the varieties of non-religious belief

By Chris Mulherin

Alain de Botton's *Religion for Atheists* may impress some as poetic prose but fails as serious thinking on religion.

I wonder how many species of atheist you know? Prompted by the Global Atheist Convention coming to Melbourne, I drew up an atheist taxonomy to make sense of the varieties of non-religious belief. Until recently my neat pigeon-holing of atheism divided my non-believing friends—with no disrespect implied—into the mad and the sad. Let me explain ...

The mad atheists typified by the so-called New Atheists, are those at the vanguard of the "God wars" currently fomented by a conflict-crazed media. These people are led by biologist and science populariser Richard Dawkins, the "high priest" of New Atheism and, like Dawkins, they are very, very angry at religion. Apart from their rage, they can be recognised by four further characteristics: their belief that religion is to blame for the world's woes; their dogma that science is the one and only road to truth; their ability to quote *Hitchhiker's Guide to the Galaxy* verbatim; and being early adopters of technology: they know that the new iPad is called the New iPad and not the iPad 3.

The sad atheists on the other hand are those who wrestle with the God question seriously. They know that the stakes are high and that without God it is notoriously difficult to make sense of the world or of human life or death or joy or justice or even, at the philosophical end of the spectrum, of truth itself. But despite the cost, the sad atheist is convinced that there is no One who might offer a well of life-giving meaning to quell our anxieties.

Such was my neat dichotomy of atheism until it was rent asunder by popular philosopher Alain de Botton. Unless you are mediaphobic you would not have missed the visit to Australia of de Botton; he received copious coverage promoting his book, *Religion for Atheists: A Non-Believer's Guide to the Uses of Religion*. And it is de Botton who has forced me to expand my taxonomy, adding another category—the glad—to the mad and the sad.

De Botton is a prime example of the remarkable levity of the glad atheist who floats through the godless life with nary a care for the substantive issues at stake. The first sentence of de Botton's book is a shot over the bows of Dawkins and Co. as well as an upset to the seriously religious: "The most boring and unproductive question one can ask of any religion is whether or not it is *true*."

For de Botton, the tragedy of atheism is that it threw out the wonderful trappings of religion with the dirty bathwater of belief in God. "Of course no religions are true in any god-given sense," he says in the second sentence of the book, after which he proceeds cheerily to ignore the question that serious thinkers, atheist and religious, have grappled with for thousands of years.

Religion for Atheists is based on the eminently wise observation that in doing away with God we lose much more than a propositional truth. Religion, and mostly the Judeo-Christian tradition, has given us the best of Western culture and we risk losing it.

Like youth, which is wasted on the young, de Botton says (in the last line of his book) that religions are "too useful, effective and intelligent to be abandoned to the religious alone." So, the urgent task lies in the scramble to hold on to religion's pearls before they are lost forever; as one wag summarised the book, "God is dead! Can I have his stuff?"

De Botton is civilised and winsome, smooth and supremely but gently confident. The contrast with Dawkins is chalk and cheese, the one crass, cranky and almost cruel in disparaging those who irritate him, the other dapper, debonair and dripping with mellifluous phrases that hint at a lack of substance.

At times *Religion for Atheists* is an incisive description of the postmodern malaise as it draws on the received wisdom of religion: it warns of "attachments to earthly status" (pp. 31–35) and reminds us of "the importance of the inner values of love and charity" and "the possibility of being happy without money." It challenges the secular world's devotion to a narrative of improvement with its "messianic faith in … science, technology and commerce" (p. 182) and it recognises the importance of confession and forgiveness (p. 55) along with the danger of "the sin of pride [which] takes over our personalities and shuts us off from those around us" (p. 35).

But *Religion for Atheists* is also patronisingly moralistic: the book is chock-full of "shoulds" and "musts," of unfounded pronouncements about what is good for humanity and what we ought to do about it. There seems no rhyme or reason for de Botton's selection from the moral supermarket shelves. By turns he's ascetic and libertarian, egalitarian and yet condescending. He applauds discipline but also license, kindness and egotism, fidelity and orgies—albeit only annually. And all in the service of aesthetic and visceral fulfilment (p. 158).

Notwithstanding his moralism, de Botton writes with style, capturing the imagination with his imagery and prose. For example:

> The incompatibility between the grandeur of our aspirations and the mean reality of our condition generates the violent disappointments which rack our days and etch themselves in lines of acrimony across our faces. (p. 181)

But despite its striking description of the symptoms of the modern malaise, *Religion for Atheists* fails to address the cause of the disease. De Botton tries to re-pot the bright flowers plucked from religion's garden into the drab flower box of atheism but he fails to realise that the flowers will not survive torn from their transcendental roots.

Even more than the Dawkins brigade, which has been criticised by foe and friend alike for its lack of clear thinking, de Botton's argument is opaque. The book's appeal is not to reason but to aesthetic sensibilities. As poetic prose this book will find some favour but as serious thinking about religion and atheism it will win few friends in its confusion of an aesthetic appreciation of faith with the hard intellectual work of making sense of religion without God.

Is de Botton really suggesting that we "rewrite the agendas for our museums so that art can begin to serve the needs of psychology"? (p. 244). Does he really think that "secular education will never succeed in reaching its potential until humanities lecturers are sent to be trained by African-American Pentecostal preachers"? (p. 131).

Are we to take seriously the following suggestions?

- In this post-God religion, architecture is central: in order to celebrate and maintain ethical norms we need to build temples to kindness, serenity, forgiveness, reflection and self-knowledge.
- A new Tate Museum would have seven levels (there is a schematic on p. 245) including galleries of suffering, compassion, fear, love and self-knowledge.
- Giorgio Armani should "run a therapy unit or a liberal arts college."
- Psychoanalytically astute travel agents will "carefully analyse our deficiencies and match us up with parts of the world which would have the power to heal us."

- And: "It is a failing of historic proportions that BMW's concern for rigour and precision has not stretched to founding a school or a political party." (pp. 291f)

While it presents as a serious proposal for post-God transcendence, *Religion for Atheists* is better described as the musings of an aesthete who dreams of re-religionising culture. But the dreams are wild and ungrounded and the book finishes up as an elegy for a fading world of religious hopes and values.

Let me finish by quoting a paragraph that captures the essence of this incongruous book. De Botton is convinced of our need for transcendent perspectives and that there are no secular alternatives to serious religious sensibilities. So, in order to invoke such feelings he turns to the stars in heaven:

> It is through their contemplation that the secular are afforded the best chance of experiencing redemptive feelings of awe. ... Whatever their value may be to science, the stars are in the end no less valuable to mankind [sic] as solutions to our megalomania, self-pity and anxiety. To answer our need to be repeatedly connected through our senses to ideas of transcendence, we should insist that a percentage of all prominently positioned television screens on public view be hooked up to live feeds from the transponders of our extraplanetary telescopes. (pp. 201f)

There in one paragraph is the essence of *Religion for Atheists*: the poetry mixed unsuccessfully with serious philosophy, the moralism, the amalgam of psychology and sociology along with a genuflexion to science and finally the banality of the idea that the transcendence of religious belief can be captured by a new law legislating that a percentage of public monitors should be tuned to Discovery Channel.

Who will enjoy this book? *Religion for Atheists* is for the *philosophe amateur* who enjoys mixing talk of the Palaeozoic age with art and history and the story of Job. With its hundred or so photos and numbered short sections, this is a boutique book; a miniature coffee table book for the chattering classes and for book clubs of the "let's never have a disagreement" type. It is poetic, beautiful at times, but not profound, suited to those who take their religion or atheism with lots of water.

Good, Evil and God

11. Genetics, evolution, cancer, suffering and God: An interview with Graeme Finlay

By Chris Mulherin

Dr Graeme Finlay is a New Zealand cell biologist investigating cancer at the University of Auckland. He is also a Christian who ponders the theological links between his cancer research and Christian thinking about suffering and the origins of humanity.

Chris Mulherin: Graeme, how did you first get involved in genetics and cancer research?
Graeme Finlay: I was raised in Asia of missionary parents. My Asian friends hugely valued education and that attitude rubbed off on me. When I returned to New Zealand as a 15-year-old, I found I could not understand Kiwi kids—I experienced culture shock with a vengeance—so it was easier just to keep to my books. I obtained my PhD in cellular immunology and was offered a job in the Auckland Cancer Society Research Lab. My task was to try to grow cells from patients' tumours in order to identify effective anti-cancer drugs. Work on anti-cancer drugs followed for 20 years, and then the group started studying the genetics of cancer cells.

The early days in the 1980s were exciting times, because people overseas were starting to identify genes that, when damaged, caused cancer. Part of this research involved a fascinating class of cancer-causing viruses called retroviruses. So I read about retroviruses in cancer, and suddenly found myself reading about evolutionary genetics. Before this, I had never been interested in biological evolution—it seemed to be a morass of controversy. But the new genetic data were so convincing; I felt that I just had to tell people about it.

So, in your explorations of cancer, you realised that cancer genetics is a model for evolutionary genetics; is that right?
Exactly. When retroviruses infect cells, they insert their little piece of genetic material, more-or-less at random, into the DNA—the genetic code—of the infected cell. The retroviral insert is then inherited by all the cells that are descended from the first infected cell. To put it another way, if all the cells in a cancer share the same retroviral insert, we can be sure that

all those cancer cells are descendants of the one original cell in which that retroviral insertion event occurred.

Now amazingly, about eight percent of the DNA that we have inherited from our ancestors is retroviral in origin. That eight percent represents about 400,000 individual inserts into our DNA code. And remember that each of those inserts has its origins in one cell which has had its DNA (including the retroviral insert) passed on down the generations.

Now if multiple species shared exactly the same retroviral inserts into their genetic code, we could be sure that all those species were descendants of the one original reproductive cell in which that insertion event occurred.

And the punchline is ...

Nearly every one of those inserted pieces of retroviral DNA in humans is shared with other primate species.

So, in effect, you are saying that the genetic revolution of the last 20 years or so offers incontrovertible evidence for the biological evolution of humans from other species? It is genetically obvious that we and the great apes have a common ancestor?

I do not wish to minimise the work done over the decades by people in other fields, but the era of comparative genomics has provided an amazing new level of precision in mapping out our genetic history. The human genome sequence was first published in 2001 and has been refined and extended ever since then. DNA is an extraordinary information-bearing molecule: it bears the instructions for our body plan. And it bears a detailed record of its own formative history. In addition to the retroviral inserts, millions of uniquely arising genetic features are shared with other species. So, yes, we know with a huge degree of confidence that we share the same ancestors with those species—not only great apes, but all primates, cats and dogs and opossums.

This is very powerful science, and, for many Christians, it might come as a threat to their faith? How did it affect you?

I was never troubled by the concept of evolution. In fact, when I started university, I decided to accept what I was told in my introductory biology courses. I accepted evolution in a lazy way—I thought it was the thing to do as a student of biology. And I did not want to base my faith on any ideas that might prove to be shaky foundations. So when I stumbled upon the genetic evidence, I did not have any anxieties—in fact, I found it to be exhilarating because the genetic data provided such compelling, lucid answers to the question of our evolutionary history. My biggest concern was that Christian opposition to evolution was so obviously wrong that I knew it would bring Christian faith into disrepute, if not contempt.

So where did you go from there?
The development of my thinking was greatly helped by exposure to Christian scholars. My professor of zoology, John Morton, made no secret of his Christian convictions. I read Donald MacKay, a British neuroscientist; he was probably the person who helped me to distinguish between evolution as *science* and evolutionism as *atheistic philosophy*.

Is this a distinction between the natural and the supernatural? A sharp divide between God's world and a natural world that gets along without God?
No; the laws of nature are not alternatives to God's action, but simply the ways we have described and codified God's action. All too often, Christians reject "natural" explanations as being "naturalistic." But if natural laws are ordained by my all-wise heavenly Father, then I should study them as expressions of his will and purpose, and I should not expect to find "gaps" in natural processes, where God needs to miraculously tweak the system.

Are you saying that what we call the laws of nature are just as much an expression of God's sovereignty and action as when an apparent "miracle" occurs?
Absolutely. Biblical creation entails that God gives being to all of physical reality, including the way by which matter and energy behave with such fruitful consistency. Take, for example, the current mystery of the origin of the first living things. As a scientist, I recognise that the mechanism of the origin of life is a legitimate question for chemists to tackle. As a Christian, I have the confidence in the wisdom of God to believe that that world is so constituted that, yes, long ago, organic molecules assembled into cells. I do not expect that a mechanism for the origin of life will be discovered in my lifetime—but if it were, I would worship the God who ordained that such a thing should be possible.

And what we call "miracles," and the Gospels call "powerful deeds" or "signs," are equally expressions of God's faithfulness representing the laws, if you like, of the new creation, for example in Acts 2:24: "It was *impossible* that death should hold him prisoner." What we now call "natural laws" are just a subset of the ways by which God acts in total faithfulness, and which will be manifested in new and wonderful ways when his purposes are fulfilled.

What about other theological challenges raised by your science?
Life has many perplexities and challenges, of course. But science has never been one of those challenges for me. In fact, science developed as a branch of Christian theology. The Greeks invented science, but they could take it only so far, because of their religious systems; the stars above the moon were gods, for example. It was the fusion of Greek physics (minus their magic, deified heroes and gods) with biblical metaphysics (one supreme faithful creator over all creation) that enabled science to develop. The journey from faith in God to the possibility of science has always had the effect of elevating science to be an integral and cherished part of my Christian worldview.

So faith in God can lead us to thinking in a sort of scientific way, to investigating the natural world?
Yes, but I should also say that, in general, I do not think the journey goes the other way: I don't think we should expect to go from science to God. Two hundred years ago, William Paley tried to demonstrate God's involvement in creation using arguments from nature, and made all sorts of problems for the church. The Intelligent Design movement has made the same mistake. To my understanding, we *first* encounter the creator in Jesus, and *then* we can look at nature and see it as God's handiwork, expressing his will, declaring his glory—for example in Psalms 8 and 19.

Given this view of God's relationship to creation, can Christians meaningfully pray that God will intervene in everyday events such as illness and weather patterns? How do you see the relationship between divine sovereignty and human (and nature's) freedom?
I have come to see that evolution is just a form of history. It has all the messiness of, say, Old Testament history. God brings his creatures into being, and places them in a lawful world. But he gives them freedom. DNA can change to produce new genetic capacities—or cancers. People can act in love—or in hatefulness. So creation is truly free, hence the random behaviour of atoms and the sinful behaviour of people. As God's people, our deepest desire should be that we and the world should reflect God's holiness and love. Hence the utter necessity of prayer. But I cannot claim to know how God responds to our prayers and works in his world. Except that, as

chapter 12 of the Gospel of John says, he draws creation to himself—he does not coerce.

So nature's "freedom" to produce cancers for example, is akin to human freedom? God doesn't force good outcomes to occur? In fact there are outcomes—in human choice and in nature's freedom, such as cruelty and cancer—that do result in evil, but God allows that space of possibilities?

Yes, I think so, but that freedom occurs within the constraints of God's faithful and wise creative action, which is so wondrously fruitful. We long for the time when the histories of matter and of Israel will come to their fulfilment in the kingdom of God. In the mystery of prayer we ask that God will move creation closer to its fulfilment—and of course we seek to align our own wills with that of God.

Another theological elephant in the science-and-Christianity room is how we understand the first chapters of Genesis. Christians of many stripes take the Bible seriously, and some would say that if evolution is true, then we can't take the stories of Adam and Eve seriously; that there is a conflict between science and faith at that point. As a conservative Christian, with a "high" view of Scripture, how do you respond?

I am but a cell biologist, not an Old Testament scholar! But my understanding of Old Testament scholarship requires me to reject a simplistic dichotomy, which says that the early chapters of Genesis are either exact science or ignorant legend. They are in fact carefully constructed stories, based on themes that pervaded the pagan Ancient Near East, but filled with astonishing new theological content. These early stories of Genesis use familiar motifs like gardens, trees or snakes, to debunk, with extraordinary audacity, the vicious gods of the empires, and to proclaim the holy, good, and redeeming God of tiny Israel. How Israel came to this revolution has to be a marvel of revelation. It leaves us with the imperative, not to argue about whether Adam was a Neolithic farmer, but to discover how we, as Yahweh's people, are to live in the face of the deepening wretchedness of twenty-first century paganism. Physical anthropology is not the point at issue—faithfulness to Yahweh is.

As we have seen, this conversation is fraught with tension for many believers as they try to reconcile traditional Christian views with

cutting-edge science. How do you advise people within the church—leaders, scientists, theologians—to tackle such questions in a healthy manner?

God is sovereign. He sustains the processes of creation—called "secondary causes" in the past—in wisdom and faithfulness. We demean the work of God if we decry the laws of nature or the mechanisms of genetics as being in some way independent of, or contrary to, the action of God. Science, as the quest to understand, should have no shackles. But technology, as the quest to control and manipulate, is another matter. We can develop medicines or bombs; more efficient modes of horticulture or trivial consumer goods; we can alleviate suffering or degrade the life-support systems of other people.

Our priority is the gospel of Jesus. Science knows nothing of repentance, justice, relationship or destiny. People formed by the sacrificial death and resurrection of Jesus need to see that they have been summonsed—elected—to serve. Western Christians must be re-orientated from protecting their privileges to bringing God's wholeness to societies and ecosystems shattered by greed and selfishness. The fields are white but the labourers are disastrously distracted.

12. Challenging Stephen Fry's diatribe against God

By Shane Clifton

Shane Clifton, who became a quadriplegic in 2010 after an accident, says that while he struggles with faith and doubt and is sympathetic with Stephen Fry's diatribe against God over the issue of suffering, he still, even in his darkest times, experiences the presence of the Spirit, which continues to give his life hope and meaning.

In a video viewed more than six million times on YouTube, Stephen Fry was asked what he'd say were he to meet God at the pearly gates. He replied: I would say to God, "Bone cancer in children, what's that all about? How dare you create a world in which there is such misery that is not our fault. It's not right. It's utterly evil. Why should I respect a capricious, mean-minded, stupid God who creates a world that is so full of injustice and pain?"

Here Fry sets out a vivid and powerful version of the atheist argument from evil. There are various ways in which theologians and philosophers have responded to the problem of pain, and while they contain substantive insight, too often philosophical theodicy talks about evil and suffering in the abstract, setting aside the existential—the personal experience of suffering that defies the impassive logic that frames abstract talk of evil. This is particularly so in public debates between atheists and theists which, especially in online comment forums, too often devolve into *ad hominem*, shallow and vitriolic attacks.

In this light, before responding to Fry, it's important that I comment on my own situation and clarify some terminology. My examination of this topic stems from an accident I had in 2010 that left me a quadriplegic. So I have some sympathy with Fry's complaint, and since the accident I have struggled with faith and doubt. Even though I mostly come out on the side of belief, I may well air a complaint or two when and if I arrive at the pearly gates.

I say this because the problem of pain is not a theoretical issue for me, but a thoroughly personal one. This doesn't make my thinking about the topic any better than anyone else's, but it does stand as a reminder that this is not principally an intellectual topic.

Good, Evil and God

The questions, "Why has this happened to me?" and, "Where are you, God?" are much more than topics for debate, but get to the heart of what it is to be human. Perhaps what distinguishes us most from other creatures is our capacity to ask, "Why?" and to try to make sense of life in the midst of both its wonders and its horrors. Because this is so, the problem of pain demands a sensitivity and generosity that seems so rarely evident in contemporary debate.

With respect to terminology, it's important to note that pain, in and of itself, is not the issue. As a person who suffers from distorted pain signals, I've learned that pain serves the purpose of warning us of danger. So the problem isn't pain per se, but ongoing meaningless pain (physical and psychological)—in other words, purposeless suffering.

There is also an obvious difference between suffering and evil. The latter references sinful human choices, whereas the former also incorporates the suffering that comes from natural causes. In citing bone cancer, Fry is arguing from the hardship of human fragility. I'll return to that issue presently, but I believe there is value in contemplating our own responsibility for the suffering endemic in human society.

The usual starting point for dealing with the problem of pain is to assert that suffering is a consequence of human sin. Whether or not this carries explanatory insight may depend upon a person's view of divine sovereignty. Those who hold to a Reformed position, understanding God's sovereignty as absolute, determining every action of the creation including human choices, cannot reference sin to absolve God of the blame for human suffering.

Alternatively, those who draw from the tradition of Augustine and Aquinas (including myself) hold that God is the primary cause of everything that is true, good and beautiful—but not of evil. Evil is understood as privation, as the absence of the true and the good, and so it is the negation of God. God doesn't cause what he explicitly condemns, and so isn't responsible for the suffering caused by human evil.

Even so, the assertion that suffering is a consequence of human sin needs to be handled with care. That evil is the negation of God, the absence of truth, goodness and reason, is important to underscore, because it reminds us that we shouldn't draw on evil to explain or justify anything, not even the problem of pain. Evil, by definition, is unreasonable and unfathomable. So, the only appropriate response to evil is protest and condemnation, as well as determined efforts to work toward healing and restoration. When a horrible evil is cited as a reason for atheism, a theist should resist replying with any suggestion that God has his reasons for causing or allowing such an evil. This is not only logically absurd, but it diminishes the horror of evil, as if there could ever be justification for rape, murder, abuse and so on.

In fact, in attempting to make this case, I am in danger of making evil an abstraction, of talking about it in theory. But evil is a concrete event that demands a concrete response.

At this point, an atheist would rightly respond, "But if God were good, why doesn't he intervene?" Similarly, some argue that if God were real (or loving), then he should have created and superintended things in some other way, so that suffering was eliminated. It is not obvious, however, what that "some other way" might be. If the freedom to choose is considered to be central to human identity and action, then liberty to pursue truth, to do good and to love is necessarily matched by the freedom to be irrational, evil and hateful. For God to eliminate free will, and hence evil choices—if that is what is meant by intervention—is to make robots of humanity.

Moving beyond the issue of freedom and sin, many formal theodicies argue that suffering (including evil) must serve a greater purpose; that although in our finitude we cannot comprehend it, God's loving purposes outweigh and overwhelm the suffering we experience. The point being made is that atheists cannot make a logical case against God based on suffering, because they aren't in a position to comprehend God's reasons for ordering the world in the way that he has. Christians have many good reasons for believing in the loving character of God—especially in the life, message and sacrifice of Jesus—and so can trust that God has a grand loving purpose for permitting suffering, one that is further supported by the promised resolution: the future new life when justice is restored and suffering done away with.

There is some truth here. Thinking about my disability, I've met quadriplegics who claim that if they could have their life over, they wouldn't change a thing; that their disability has enriched their life in irreplaceable ways. Many in the deaf community have come to a similar conclusion, understanding their seeming impairment as a gift. It is probably true that most people who live with a disability would be able to identify some good that has resulted from their impairment.

In my short time of living with quadriplegia, I've experienced staggering generosity, deep love, heartfelt compassion, courageous determination and exemplary care. Indeed, depending upon how we respond, a case can be made that sickness, injury and disability (whatever their cause) enrich society, since almost every human virtue arises as a response to hardship.

Even so, there is a world of difference between the assertion that good can come from hardship, and that suffering is necessary for the good.

In terms of theodicy, the problem with the greater good position, especially with reference to evil, is that it diminishes suffering. So, for example, if God (or a society that practises sacrifice) causes the abuse and suffering of a child, any good that comes from that action is corrupted. No matter what the good, or the glories of a future compensation, it can't be made sense of, and it isn't worth the price that is paid. This insight shouldn't be up for debate—it's an attitude that atheists and theists should share. In drawing on evil for the sake of argument we underplay its significance, and are in danger of passivity, especially if we try to justify it. As I've already said, evil, once recognised, can only be condemned and resisted.

Still, you might ask, if God exists, why create a world in which evil is possible? And what about suffering that results from natural causes? To respond to both questions requires that we face up to the freedom, finitude and fragility that are part and parcel of what it is to be human.

I've searched high and low for an explanation for my quadriplegia, and tried to imagine the purposes of God in my situation. But ultimately, I'm a quadriplegic because to be human is to be subject to the vulnerabilities and frailties of finite life. To be human—a creature of the earth—is to be born, to grow, to break down and to die, to be limited in power, strength and knowledge, to be fragile and vulnerable, to be constituted by DNA and imperfect genes, and made up of bones that flex and break, muscles that tear, and blood that spills. It is to have the freedom to think and imagine and to make choices, but also to make mistakes and to live with regret, and to build a life in families and communities, as well as to suffer the mistakes and regrets of others—and sometimes to be subject to horrible evil and unfathomable suffering.

In his excellent book *A Public God*, Neil Ormerod responds to the idea that God should have created a world without suffering by observing that, "To repudiate the conditions from which we have emerged is to repudiate our own existence." Andrew Gleason, in *A Frightening Love: Recasting the Problem of Evil*, similarly notes, "There is an absurdity in putting an end to human life to spare us the suffering it involves."

One of the problems of modern society, with all of its medical and technological wonders, is the implicit demand that we should live forever in perfect health. We keep our dead and dying out of sight, we abort babies that don't match our ideals of normalcy, we worship a photoshopped image of beauty, and in consequence, suffering, disability and fragility come as a complete and utter shock. We just don't know what to do with them.

Why then, you might be thinking, do you believe in God? While I really need longer than one paragraph to answer that question, let me conclude as I started, by noting that I empathise with Fry in his complaint to God. Clearly for him, suffering is not just an intellectual problem (although, if God does not exist, I wonder with whom he's really angry?). The writers of Scripture, I think, would also be sympathetic; consider Job, or the psalmists, or Jesus himself: "My God my God, why have you forsaken me?"

Nevertheless, in addition to hating this bloody, broken body, I have also come to appreciate that life is wondrous, and sometimes even sacred. This is not to weigh the good against the bad, but rather, at one and the same time, to see the wonder and tragedy of life, and to identify God in both.

This position isn't irrational or anti-science, but instead understands God as the ground (the primary cause) of the laws of nature and the beauty of the natural world. So, while God isn't the cause of evil, I have found that the experience of the Spirit, sensed in the darkest of times, has provided hope and given my life meaning.

This chapter is a slightly revised version of an article which was first published on the ABC Religion and Ethics online journal on 26 February 2015.

13. Wrestling with evil: Can a "reckless" creator God be good?

By Stephen Ames

Is the kind of world that the sciences know and describe, and which includes tsunamis and earthquakes, one which a loving God would create? Perhaps God could be described as "reckless" and if so, could God be truly good?

I find there are at least two kinds of answers to objections to belief in God. The first kind of answer shows that an objection to belief in God doesn't hold. A simple example is the way people object to belief in God because the Genesis account of God creating the world is, to them, clearly out of date compared to what science tells us. It is possible to show that this sort of thinking confuses the relation between the Bible and science and is unaware of the important history of science and religion, especially Galileo's "two books" principle. This kind of answer defends belief in God by showing that an objection does not carry weight. This is strengthening for believers and may help the questioner think again.

This kind of answer may go further and also seek to draw the questioner into thinking more deeply about the God to which they are objecting. This can be done by saying something like "Tell me about the God you don't believe in." Then, when an answer is supplied, the response may well be, "Oh, I too don't believe in that kind of God." The conversation may then continue as the believer gives an account of the God he or she does believe in, hopefully in a way that engages the questioner and appreciates the question from the standpoint of the questioner and their perspective on the world.

Objections based on suffering and death

Let's turn to a specific example. For many people the natural processes by which the earth was formed and now operates, and by which life has evolved, produce so much suffering and death that it seems obviously not good. Surely this could not be a universe created by a wholly good God? The scientific view of the natural world, from this perspective, is quite different from what you would expect if the universe were freely created and sustained by a God who is all good, all powerful and all knowing. This objection is often known as "the problem of natural evil."

The questioner is making the point that you would expect a good God to create a different kind of universe: one which has the positive values of evolution but without the suffering, pain and death involved. This way of understanding God and the world presupposes an argument *from* an idea of God *to* our expecting such a God to create a different kind of world from the actual world. For example, the questioner might say that a perfectly good God would create a universe in its perfect end state, thereby fulfilling God's purpose; there would be no process leading to that end state. Clearly that is not the universe science tells us about, nor the one the Bible tells us about. So, the questioner would continue, the problem of natural evil justifies rejecting the idea of the all-good, all-powerful and all-knowing God who has created this universe for some purpose.

In response we can and need to go beyond the first kind of answer to the sceptic to a second kind of answer.

What kind of universe would you expect a wholly good God to create?

My second kind of answer aims to describe what kind of universe we should expect God to create, hopefully in a way that is consistent with an evolving world, where suffering and death are crucial to producing life.

Let's turn to thirteenth-century theologian St Thomas Aquinas who asked: does God create things with their own real powers or is God the only power in the world? Is it the fire that keeps you warm or is it God in the fire who keeps you warm? Aquinas argued that we should prefer the first view, where things have their own powers, because otherwise creatures would have a pointless existence. For Aquinas, the idea that God is the only power in the world implies a lack of power in the creator to create things with their own powers and we should not attribute such a lack to God. Why? Because when we speak about God as wholly good, all-knowing and all-powerful, we are speaking in superlative terms.

So, it is preferable to say that God creates things with their own powers that are capable of doing good for other things. For Aquinas, God creates things in such a way that things have the dignity of also being causes, rather than, so I would add, the indignity of not having the power of causation. On similar reasoning we should say that God maximises these features of creation (e.g., creatures with their own real powers, good in themselves and good in what they can do), rather than minimises them.

We should therefore prefer to say that this God creates a life-producing universe, which is better than producing only an inert universe, or a chaotic universe, or a merely mechanically interactive universe. Therefore, we should expect that things create other things and that overall creation creates itself as much as possible as a life-producing universe. Of course, this is easily extended to a life-producing universe that produces intelligent life, whether like us or otherwise. On this view, what is of value to God is creating creatures that are also co-creators.

This discussion gives rise to several questions. For example, Why is this a *better* type of creation? Briefly, the answer is that this "recklessly creative" universe is more like the life-giving God who calls into existence what does not exist, whose Spirit broods over creation, who raises the crucified Jesus by this Spirit—"the Lord and giver of life."

By creating a life-producing universe God maximises the possibilities of creatures with their own real God-given powers, as co-creators of life, including intelligent life. Creatures are "co-creators" because by them God brings new things into existence. But only God creates *ex nihilo*—out of nothing. This is an expression of one thing that is of value to God—creatures as co-creators.

God's freedom means science cannot discern God's purposes

Here we have some broad themes concerning what the universe created by God would be like. It does not provide specific detail of what the universe would look like; for example, it doesn't say how the God-given powers of created things would actually work. This is something theology recognises that it cannot do. It cannot work *from* the idea of God to the specific detail of God's creation. Why not? This is so because, theologically, creation is the *free action* of an all-wise and sovereign God. Because God is free, reason alone cannot specify what God freely chooses.

To find out what in particular God has freely and intelligently created, we have to go and observe; we have to inquire with our senses, our imagination and reason—in short, we have to use the methods of the natural sciences. They are a testimony to God's freedom and rationality. We must inquire. And when we do we find a life-producing universe, with creatures as co-creators, and this whole process maximises these possibilities from the utterly small sub-atomic particles to the utterly vast galaxies, producing heavy metals in the belly of exploding stars seeding the universe with materials needed for life.

One aspect of this recklessly creative universe is captured in what we call natural evil. We discover that the natural processes, with their God-given powers, which produce the many good things in the world, also produce earthquakes, tsunamis and floods. One powerful example is the processes of our continents riding on tectonic plates, and, in counteraction with erosional forces, generating and regenerating landscapes and seas. These same tectonic processes contribute to underwater earthquakes in locations where an oceanic plate is being sucked down into the earth's mantle. The earthquake leads to a tsunami ... however, it is also possible that life in the form of micro-organisms first came into existence in these same processes, in the chemically rich, superhot broth spewing from vents in the sea bed.

So, in this second form of argument, we find a universe in which what is of value to God is realised at different levels—creatures are co-creators up and down the tree of life—and we see that we have come to this view without any *necessary* knowledge of God's purpose in creation. This suggests that the scientific study of the world shows us what is of value to God but it does not show us God's purpose in creation.

God's purposes

So what shall we say is God's purpose in creating? And how is that purpose to be understood in relation to a scientific picture of the universe which appears to deny any purpose—for example the universe endlessly expanding into a dark, cold, undifferentiated sea of energy known as the "big freeze"? Part of the answer is that these "entropic" processes of increasing disorder are connected to how more complex things are produced by our natural processes. Theologically, they are part of the way creatures are co-creators and so express what is of value to God—all within God's purpose and ultimate governance of the created world.

In the philosophy of Thomas Aquinas, governance is for the sake of the governed to bring them to their perfection. A higher perfection and a higher purpose go together. I cannot say what is the "highest" purpose God might have in creation, but I do think that a wholly good God would invite the creation into a loving and just relationship with God. Any idea of God's purpose that did not include this purpose would be a lesser purpose so we should prefer to say that God's purpose would include drawing the whole creation into a loving and just relationship with God. But how does that purpose fit with the scientific prospect of a "dead-end" universe? Such a relationship would not be the equivalent of a "one night stand" but rather an invitation to explore the fullness of God. It would require the transformation of the dead-end universe in the resurrection of the dead universe—something that we see anticipated in the bodily resurrection of Jesus.

Even with this hope many people would ask whether a truly good God would be so "reckless" as to create this kind of world if it means that such a world produces tsunamis and terrible genetic disorders? If God is "reckless," can we still say that God is wholly good?

The "reckless" goodness of God

For Christians, this "recklessness" of God is tied closely to God's goodness and love, pointing us towards an answer to this question.

An example: In the parable of the Prodigal Son the son outrageously asks his father for half the family inheritance. Culturally understood, he wished his father were dead. Even more outrageously the father in the parable gives his younger son half the heritage and allows him to leave. Many would say this was a reckless act, showing an extra-ordinary sort of love for his son. The father in the parable gives us a glimpse of God "recklessly" allowing freedom to his creatures.

Another example: In Genesis 1:26-28, and in Psalm 8, we see the "recklessness" of God in creating humankind with such extraordinary powers (something the gods of other nations restricted to royalty alone) and the freedom to use them for good, but with the risk of also using them otherwise.

A final example of the "reckless" goodness of God is seen in the cross. Having given created beings real powers, these powers are violently visited upon the triune God who submits to their evil on a cross, in order to absorb, limit, transform and overcome evil, in the life, death and resurrection of the incarnate Son of the Father.

Here are sound reasons for thinking of the truly good God in a way that supports the idea that this God, the creator of the universe, would be so "reckless", even "foolish," in creating this universe, in liberally giving to created things real powers and therewith the dignity of being co-creators, indeed maximising this co-creating in a life-producing universe. This God is the "reckless" creator of this life-producing universe, whose foolishness is wiser than human wisdom, who is also the "reckless" redeemer, loving, wholly good—indeed that good of which there is none greater—whose glory will be revealed in the renewal of the creation itself.

14. After disaster, allow space for hard questions

By Steven Ogden

It is important to give people permission to raise and discuss the hard questions about our faith, especially after natural disasters and during Lent.

My wife calls me *the kiss of death*. This term of endearment came out of our experience of attending wedding receptions. As a priest, I have the opportunity of participating in many of these festive events. Typically, as we enter the reception room, the first thing we do is look for the display board, which reveals the unfathomable mystery of the seating arrangements. Invariably we are placed on Table 17; the miscellaneous table.

The table is situated at the back of the room, next to the table where the three-piece wedding band adjourns for a beer and a bowl of fries. I usually wear a clerical collar with my dinner suit. This means that as we meander across the reception room toward our table, the guests freeze with dread suspecting that the priest is coming to sit at their table. As we pass by successive tables their sense of relief is palpable, and in some cases audible, as guests can be heard to mutter what sounds like a profanity but could be a prayer, "Thank God. He's not sitting here."

We arrive at Table 17. Our table party consists of Uncle Bob a retired engineer, Trevor the family accountant, Shona the bride's long-lost girlfriend and an aloof cousin from Manchester. They are polite but uncomfortable, as they try in vain to mask their misfortune. They soon relax however, as we share in the table banter and indulge in a glass of wine. Without fail, the guests proceed in turn to announce "I am not religious" or "I'm spiritual, but I don't go to church." Nevertheless, we end up having a wonderful evening full of animated conversations about families and work, success and failure, life and death. When the evening concludes, our companions say in all sincerity that they enjoyed our company and that they were surprised to discover we were perfectly normal.

The term *the kiss of death* is a form of gallows humour. It has its lighter side, but the wedding experience is a reminder that the public image of the church (its brand) leaves a lot to be desired. I know the church is doing great things in the world, but for much of the world, religion is an embarrassing, even vexing institution. One of the stumbling blocks is *the God problem*, which can be summed up by the question, "If God is a loving and powerful God, why do bad things happen?" This is what theologians refer to as the problem of theodicy or the problem of *God's justice*. However, the problem is not about us coming to terms with God's justice, but rather, how and why our understanding of God has changed in the light of the twenty-first century experience of suffering.

My grandfather died when I was nine years old. I did not attend the funeral, which was not unusual then, but I recall with remarkable clarity my parents returning from the funeral. What struck me then was a rogue thought, seemingly from nowhere, saying "so much for God." God's silence was deafening. In the end we owe it to ourselves, and the guests at Table 17, to be honest about suffering.

After a disaster, we look for a word or a sign to galvanise our thoughts and emotions, so that we can respond immediately to pressing needs. In due course, however, there will be time for expressing a sense of loss and asking the hard questions. Pastorally, it is important to give people permission to raise and discuss these questions. Theologically, it is important to ask not only where God is but also which God are we talking about?

Before the Enlightenment, and prior to the Renaissance, a pre-modern understanding of God was operative. In this view, God was understood primarily as a heavenly or transcendent sovereign being, bringing order and purpose to a messy and menacing world. And of course, how else would pre-moderns understand God? This theology is expressed superbly in the architecture of Gothic cathedrals, where vaulted ceilings express the view that God is in charge and everything has its place. This was reassuring then, but is it convincing now?

If two world wars, Auschwitz, and recent natural disasters have taught us anything, it is that a pre-modern view of God is not working for our time and circumstance. This is not a criticism of God per se. God is good, loving and faithful. But the way we construe the significance of the mystery of God changes over time. In other words, the experience of the incomprehensible mystery of God is a divine constant, but our perceptions of divine mystery change. This means somebody else's pre-modern understanding of God does not necessarily speak to our postmodern experience. It may well be a source of comfort or a benchmark for faith but if we are going to reclaim hope, then we have to translate the meaning of our experience into twenty-first century theological terms.

After disaster, allow space for hard questions

There is ground for hope, however, which can be embraced in the season of Lent and the experience of Holy Week. This involves us bringing our raw experience and hard questions to the cross. For example, historically the pre-modern God had been portrayed as perfect in the sense of being separate from or immune to suffering. But if the cross has taught us anything, it is that God is not indifferent to our plight. On the contrary, God feels for us and with us in our suffering. Instead of thinking of God as impervious to human anguish, we take heart in the pathos and compassion of God. Instead of thinking of God as in control, we find hope in God's vulnerability.

All this represents a major shift in thinking from God as a power over us to God as grace among us. And this God, this Christ-like God, speaks to a postmodern experience of suffering. In this light, Lent and Holy Week are replete with opportunities for people to honour their experience and sit with the difficult questions. To begin, Ash Wednesday with the imposition of ashes speaks poignantly about our humanity and the complexities of the human predicament. In the end, Good Friday speaks passionately about a suffering God who is with us; even when we feel abandoned.

The experience at a wedding, with our companions on Table 17, is a reminder that most people know the pain of the absent God. But there is a word of hope. Out of the shadow of the cross, new light has come. For us, Jesus is a potent and impassioned symbol of this hope. Suffering is a reality. There are no simple answers. But there is the hope that the present suffering will not have the final say.

We cannot go back and repair the past, but with God's grace we can begin again with renewed heart and spirit. Our source of hope is found in a God who is a consoling yet energising presence, who enables us to survive, thrive and begin again. That's the meaning of the Cross. That's the essence of the ministry of Jesus and life in a faith community: God is with us. And because God is with us, we can grieve, we can ask the hard questions and we can hope.

15. Droughts and flooding rains: Disasters are part of the planet's functioning

By Jonathan Clarke

While cataclysmic processes are part of the normal functioning of the earth, the suffering they cause indicates that creation is not yet finished.

Floods, bushfires, storms, earthquakes and volcanic eruptions are frequent visitors to our news streams, and with good reason. These events can kill thousands, destroy communities and bring down societies. Many of the readers of this chapter will have experience of bushfires, a major feature of the Australian landscape, and perhaps other events as well. Such events can be traumatic to those who survive them and historically have made some question the goodness and indeed the existence of God.

I am a geologist with 40 years' experience studying the history of our planet and understanding how it works. I have studied five continents, three oceans and two planets (Earth and Mars). I worked in universities teaching and researching, for industry in resource exploration and extraction, and government agencies in environmental management, especially water. I do not claim to have specific answers to those troubled by such events, however I hope my reflections, based on my experience, may provide helpful tools for those who want to think further on them. As a geologist I will confine myself to physical events we call disasters—floods, fires, eruptions, earthquakes, tsunamis and the like—and not the biological disasters such as epidemic, or more general questions such as "Why does disease exist?"

In thinking about natural "events" it is important to separate the process and the consequence. There is no difference between a volcanic eruption on an uninhabited island from one on an inhabited one. The first is an interesting phenomenon, of interest mainly to scientists, worth perhaps a brief "filler" on the news. The other is potentially a disaster, with lives lost, and property and infrastructure destroyed that may dominate news coverage for days. While catastrophic events might be bad news to ecosystems, for example forest fires in uninhabited areas, red tides in coastal zones, or even the extinction of the dinosaurs after an asteroid impact, they are part of the normal functioning of creation.

Droughts and flooding rains: Disasters are part of the planet's functioning

Therefore we can say that there are no natural disasters, only human ones with natural causes. As with all human disasters, these are then driven by a combination of human ignorance, folly and injustice. People may not have known the volcano was active, they may have chosen to ignore warnings, they may have been forced to live in dangerous areas through injustice, or decided to live there because of greed. Others may have made a rational choice to balance the risks against the gains.

Cataclysmic processes are as much a part of a functional and functioning creation as the less spectacular processes we consider normal, such as rain, the tides, normal river flow, day and night. Without volcanoes we would not have an atmosphere, oceans, fertile soil and many mineral deposits. Earthquakes are the consequence of plate tectonics, without which we would not have continents, mountains and the mineral deposits found in fault zones. Floods recharge aquifers, refill wetlands, replenish floodplain soils and wash nutrients into coastal zones, as well as cause loss of life and property. Bush fires are destructive, but also the drivers of forest health and biodiversity.

A consistent theology of creation is important. If we thank God for the beauties and bounty of creation we must also thank him for the processes that he uses to provide them, including the cataclysmic ones. The Psalms refer to storms as God's judgements (83), but also signs of God's majesty (107, 148), and in Job God speaks from the storm. Tourists flock in their millions to places of great scenic beauty like Yellowstone, Santori, Taupo and Toba, even though they are the result of geologically (and sometimes historically) recent super volcanoes. Even Dorothea Mackellar's famous poem recognises that the Australian landscape owes its beauty to droughts and flooding rains, and its splendour and terror are closely linked.

Perhaps part of our difficulty lies in our misreading of Scripture. We have been so wrapped up in thinking that God created a "perfect" world that we forget the Bible nowhere says this. We are told in Genesis 1 that God was pleased with what he saw and declared it good. The goodness here is the goodness of function, just as we would declare each stage towards the completion of a house as good, even though it is still a building site. The declaration that creation is "very good" happens on the seventh day, when the world is functionally complete, just as a house is "very good" when it is complete.

This brings us to a second misunderstanding. Creation is not yet finished. The goal of creation is not simply a functional world, but a new creation. God is not yet at rest, the seventh day of creation is still to come. The perfection that some expect of our world will be realised only in the new creation, not the present one. Until that comes, we are living on a building site, a good one certainly, but still with all the hazards and risks that a building site brings.

In Romans 8 we find the creation being spoken of as metaphorically frustrated at its present state. This might be analogous to how occupants of an unfinished house are also frustrated. But it also speaks of creation for the revelation of God's children. The whole creation will one day share the glorious freedom of God's children. The present sufferings are like the pains of the birthday of the new creation. This means that creation's hope, just like ours, is based on Jesus and the cross. It is why the most famous verse in the Bible is about God's love, not simply for the world, but the entire cosmos, being expressed in the gift of his Son.

What does this mean for us living on a building site, experiencing the birth pains of a new creation? I suggest four things.

First, we should stop thinking of ourselves as being passengers in creation, but rather crew and builders. Humanity is part of the process of bringing about the new creation.

Second, we need to think realistically about the world, and not have unrealistic expectations about creation.

Third, we should stop thinking of humanity as separate from the rest of creation; we need to live in it wisely and with respect. However, unlike the rest of creation (as far as we know) humanity is uniquely gifted with agency and moral accountability. Therefore we can, and in many cases should, preserve species and ecosystems under threat from natural cataclysms, and can and should control floods and bushfires, but in a way that does not bring unintended consequences. We can and should farm and use natural resources, but always remembering that the creation has value, declares God's glory and is worth being included in God's salvation plan. It is not merely stage scenery for the drama of our salvation, but shares our hope.

Fourthly, we should, along with the Psalmists and the writer of Job, when contemplating the terror and majesty of creation, respond with awe and worship.

16. Natural disasters not a result of the fall: An interview with Bob White

By Chris Mulherin

Volcanoes, earthquakes and floods are not the result of a fallen world, argues Bob White in a conversation with Chris Mulherin.

Chris Mulherin: Bob, tell us why you set up the Faraday Institute for Science and Religion?
Bob White: I helped to set it up 11 years ago with Cambridge biologist, Denis Alexander. We started it because we were concerned that working scientists had shut their minds to the whole idea that God could exist or was in any way relevant to them. So we wanted to reach our tribe of scientists with some of the truths of the gospel or even just to say that it was reasonable to hold Christian faith.

We didn't set up the institute in a theology department, which is where most such institutes in the world are based, because most scientists would say, "Well, theologians would think that, wouldn't they?" And they wouldn't go anywhere near it. So we set it up in a neutral space in a college in Cambridge, and we specifically only hold to mainstream scientific positions. We do not proselytise and we do not campaign on behalf of anything. We're just trying to say to our colleagues, "It is reasonable to hold religious faith and it helps explain the world a lot better than just science on its own."

You also aim at non-scientists?
Yes: many groups invite us to come and talk to them: church groups, or men's breakfasts or university Christian unions. One group that we're purposefully trying to reach are ordinands, because these are the people who will be standing in pulpits in 20 years' time. By and large they are often pretty uninformed about science generally, so we'd love to educate them a bit; we run a course for the ordinands in Cambridge and they love it.

Now, you're a volcanologist ...
Geophysicist, actually.

Volcanologist sounds more dramatic ... You go where other men and women fear to tread, don't you? You tell the story of a recent eruption in Iceland with a 100-km exclusion zone around it and only you and a few

others approached the volcano as it spewed lava 150 metres into the air. Now, why would you do that?

Because volcanoes are exciting research topics. Actually these days most people who die in volcanic eruptions are volcanologists, because when it starts to erupt they rush towards it, while everybody else rushes away. This particular one was in a very remote area, so didn't hurt anyone or damage anything.

What can we learn as Christians from volcanoes, floods, tsunamis, earthquakes? These "acts of God"—at least that's the way the insurance people write them up—are they "natural evil"?

No, I don't think they're natural evil. Even though they can cause terrible devastation and loss of life they are also essential for life. Floods bring nutrient-rich soils down into the flood-plains, where we grow crops. Earthquakes enable mountains to grow, and then the mountains get eroded and bring those same soils down the rivers. Volcanoes are the main source of nutrients from the deep earth, bringing them up onto the surface of the earth. They're all parts of what make this world a very fertile place for humans to live on. If it weren't for those we wouldn't be here.

So you don't buy that argument that says, "Well, once sin came into the world we also got tsunamis, volcanoes, floods, firestorms ... "?

No, I don't buy that because we know full well that there were earthquakes, floods and volcanoes for eons before the fall of humans who first sinned and rebelled against God.

Now you're not a young earth person ...

No; the earth has been around 4,560 million years. The universe is about 13.8 billion years, so quite a lot longer—about three times older.

And there's a continuity of geological progression? Is that what we see over the last four billion years?

Oh, yes, we can see the rocks. I've got one in my cabinet at home ... the other day I was showing people the oldest known rock in the world and the youngest known rock, a piece of volcanic rock I'd picked up from the Icelandic volcano, so it was only a few days old. The oldest known rock is 4,050 million years old, which is very close to the beginning of the earth.

So, tell me, how does your understanding of the fall of humans six or eight thousand years ago fit with an evolutionary story of modern

Natural disasters not a result of the fall: An interview with Bob White

humans having been around for a hundred or two hundred thousand years?
Clearly we are animals, because we share our DNA, 95 percent of it, with chimpanzees. I think we share 80 percent of it with a banana, actually. And the evolutionary story is one of the most powerful descriptors of the interrelationships between organic life, so I have no reason to doubt it.

But the question is, what changed us from being animals to being humans made in the image of God? And I think that issue of being in the image of God is the key issue, because that's what the Bible says is the difference between us and animals; nothing else is made in the image of God.

The way I square that is that at some point God breathed awareness of himself into some hominids and turned them into humans made in the image of God. You can ask, "What does being made in the image of God mean?" There's many things ... I think a deep concern for justice is also one of those things. I'm always struck by the issue of justice because that is so deeply built into us. And clearly it's deeply built into God's nature as well, because the whole story of the Bible is about God's justice.

There are many who would disagree ... I wonder if you'd like to comment on the movement that we've seen in the last decade or so, the so-called New Atheism. What do you make of science and faith conversations in the English-speaking world, and where does the New Atheism fit in?
Yes, it's interesting, isn't it, that the New Atheists like Richard Dawkins and so on, are so violently opposed to something they say they don't believe in. Why are they so worried about religion if they think it's meaningless? And yet they clearly are. I think they suspect there's something to it, after all.

Actually, I think the New Atheists are helpful in a way because they bring this debate into the public domain. But it is a bit sad to see people like Richard Dawkins railing against his version of Christianity; they're railing against a straw man that is not the Christianity that I believe in.

What about academics who are not Christians? What is their attitude to Christianity?
Well, if you ask most of them, they're completely apathetic; they don't really care, I suspect. There are a few who rail against it and think it should have no place in a modern university. I'm quite glad, at least in Britain, that we don't have Christian universities. I think it's good to have Christians embedded in a normal university, just to stand amidst everybody else. And

they can argue the case for Christianity on the strength of the evidence, which is a very strong case.

I'd like to return to the question of natural evil. In your talks here in Australia you transferred a lot of the blame to human beings.

Yes. I'm not sure that the term "natural evil" is a helpful term, actually. Partly because "nature" is a sort of shorthand for this world we see around us. Nature is created by God: God is immanent in it, sustaining it moment by moment. So it's sort of a contradiction in terms, almost, to talk about natural evil.

As if God were not involved?

Yes, exactly. As if you've separated him out from nature. Time and time again you can see some natural process, like a volcanic eruption or an earthquake or flood, turned into a disaster by the actions of humans. Now, that doesn't account for every single death but it can account for the huge majority.

Could you give us an example?

There was an earthquake in California, at Loma Prieta in 1989. A magnitude 7 earthquake that killed 57 people. Many of us will remember another earthquake seven years ago in Haiti. It was an identical earthquake but it killed 230,000 people.

Most people died because they lived in ramshackle buildings made of concrete that just collapsed on top of them. So they died unnecessarily: we know how to build buildings that don't fall down, but Haiti had suffered decades of misrule and corruption and it's the poorest nation in the western hemisphere.

So you have to say, "Whose fault is it they died?" It wasn't their individual fault, but it was human factors that led to them living in such circumstances. And time after time the people who die in disasters are the poor and the disadvantaged or the infirm.

And so—we were talking about justice earlier—that really does seem unjust, doesn't it? And you can see the pattern repeated in disaster after disaster: there is human agency at work at some level which turns what is of itself a good process into a disaster.

Finally, what do you say about the so-called conflict between science and faith?

Although science and Christian faith have never presented any major intellectual conflict for me, it is clear that in our culture there is an often unspoken assumption that they are in fact opposed to each another. Sometimes Christians contribute to this distrust of science and retreat into statements that the Bible as interpreted literally is the only truth. The implication is that the science must be wrong.

Science is a secular activity insofar as its very strength lies in not appealing to any external causes (such as divine activity). So scientific theories can be understood in the same way by atheists, Buddhists or Christians and work equally well in Beijing, Brisbane or Budapest.

But both scientists and Christians believe that there is an underlying reality to be found; that some things are true and others palpably untrue; and that we can distinguish between them in statements that apply to all people and for all time.

17. Why does God allow suffering?

By Roland Ashby

Terrible diseases such as cancer, and natural disasters such as earthquakes and tsunamis understandably generate a great deal of soul searching among believers about why God allows pain, suffering, evil and death to exist in the world. And this question is perhaps the greatest stumbling block to faith for many outside the church.

If God is perfect, all-powerful and all-loving, why does he allow evil and suffering to exist?
Christianity has traditionally advanced various arguments to answer this question. The Free Will Defence, for example, argues that the price for creating creatures free to love and to do good, is that they are free also not to love and do good. "God is not the puppet master of either men or matter," Anglican priest and physicist Sir John Polkinghorne has asserted.
Such a defence may explain the existence of moral evil, the evil or suffering caused or inflicted by human beings, but what about natural evil such as cancer or a tsunami? Here, too, according to some theologians, it is a question of freedom. In the case of cancer it is the necessary freedom of biological cells to mutate as part of the processes of physical evolution which can also lead to the mutation of cells called cancer. And the same principle applies to natural disasters. The earth must be free to behave like a planet—including the horror of earthquakes and tsunamis—if human beings are also to enjoy the stable physical conditions which make life possible and pleasant.
The Catholic philosopher Peter Coghlan has made a similar point ("Confronting This Horror," *The Age*, 8 January 2005). Suffering caused by natural events, he says, "is a necessary feature of our being bodily creatures capable of thought and compassion. We cannot enjoy the beach at Phuket without the kind of bodies we have; but those bodies make us vulnerable when a tsunami strikes. We grieve over the death of our loved ones; but without that suffering there can be no compassion for those who mourn."
Suffering is a necessary part of loving, as the former Religion Editor of *The Age*, Barney Zwartz, reminded us ("God Alone Knows Why There Is Suffering On Earth," *The Age*, 3 January 2005) when he quoted theologian

Why does God allow suffering?

Nicholas Wolsterstorff: "Suffering is the meaning of our world. For love is the meaning. And love suffers."

Because of the freedom given to creatures and to matter, God's power, some theologians argue, is necessarily limited to exercising influence and persuasion from within the process. To exercise absolute power would remove the freedom. God is not indifferent or impassive in the face of suffering, such "process" theologians contend, indeed he is "a fellow sufferer who understands," but he cannot coerce his creatures or matter "to obey the divine will or purpose for it."

Such arguments may appear reasonable but what comfort do they offer when suffering strikes?

The case of that great apologist for the Christian faith, C. S. Lewis, is well known. The confident writer of *The Problem of Pain* is a very different man in *A Grief Observed*—his reflection on the devastating experience of the loss of his wife to cancer. An academic discussion of the problem had been superseded by a baring of the soul. He now realised that "you never know how much you really believe anything until its truth or falsehood becomes a matter of life and death to you."

For many, no doubt, whose lives have been devastated by illness, or a tsunami or earthquake, there is a shattering sense that God is absent, and even a sense that they have been abandoned by God. Lewis experienced a similar sense of absence and abandonment: "Go to him when your need is desperate, when all other help is vain, and what do you find? A door slammed in your face, and a sound of bolting and double bolting on the inside. After that, silence."

This sense of absence and abandonment was also "lived through" by Jesus. The question "My God, my God, why hast thou forsaken me?," the first words of Psalm 22, which were recalled by Jesus in his final dying moments, is perhaps the most poignant expression of suffering that we know. It describes that sense of complete abandonment, helplessness, powerlessness and vulnerability experienced by millions of human beings across the ages, when they have suffered as a result of disease, famine, war, oppression, cruelty or natural disaster.

The question is also extremely significant for Christians—not only because it affirms that the problem of evil and suffering is a practical, existential and ontological problem for humankind, but also because the words are spoken by the Son of God. Suffering, it seems, is not just a condition of being human, it's also a condition of being God!

Good, Evil and God

In his book *Night*, Elie Wiesel tells the story of the death camp Auschwitz. In one passage, a young boy is hanged for not keeping one of the camp rules. As his body dangled from the rope, Wiesel was asked by someone, "Where is God now?" and a voice within him replied, "Where is he? He is here—he is here hanging on this gallows."

Thus in the cross do we see one of the great paradoxes and mysteries of Christianity—that God may be all powerful, but his power, it seems, can only be exercised in utter powerlessness.

Is it credible that God, through his son Jesus, has become "a fellow sufferer who understands"? At first glance, this claim seems not to adequately answer Ivan Karamazov's damning indictment, in Dostoevsky's book *The Brothers Karamazov*, of the God who allows the suffering of innocent children. Ivan is rightly outraged at the senseless and cruel suffering of small children, and cannot accept this may be part of the price of "eternal harmony" or "making men happy in the end, of giving them peace and contentment at last."

This theme of God's apparent indifference to suffering is also seen in the book of Job, where God seemingly allows a righteous man to suffer for no reason.

But for Jürgen Moltmann, for whom the problem of evil and suffering became real when he was a POW, the answer to Ivan is to be found in the death and resurrection of Christ. Indeed, it is this death, the death of an innocent sufferer, which for him lies at the very heart of the Christian faith. It is at this very moment that the incarnation becomes complete: God enters into the utmost human desolation, in a supreme act of "loving solidarity with all those who suffer apparently abandoned by God." God willingly and lovingly comes alongside to share the pain of those who suffer, and "voluntarily identifies with them, takes their side and shares their fate."

However, for Moltmann, in this act God is not only identifying with the "Godforsaken" in their suffering and abandonment. He is also, with Ivan, protesting against such suffering and injustice, and in the resurrection, holds out a hope and promise that it will be overcome. "This divine promise gives no explanation of suffering, but hope for liberation from suffering." Such a hope is also significant in Karl Barth's thinking about evil. For him, the believer can find strength and confidence from the belief that God's grace will ultimately triumph over all evil. But such consolation is eschatological. How are believers to confront the question of the present reality of evil and suffering in their lives?

Why does God allow suffering?

Although Dorothy Soelle sympathises with Ivan, it is to his brother Alyosha that Christians, she says, should look for inspiration. While Ivan rejects God because he demands both a complete intellectual answer to the problem of evil and an end to suffering, Alyosha chooses the path of discipleship without a clear answer to the problem. "Where Ivan turns to heaven in accusation, Alyosha pursues an earthly *Imitatio Christi* that involves solidarity with those who suffer."

It is Alyosha who has learnt, in Bonhoeffer's words "to see the great events of world history from below, from the perspective of the outcast, the suspect, the maltreated, the powerless, the oppressed, the reviled." To live in solidarity with those who suffer, Soelle maintains, is to live in solidarity with "the suffering Christ who represents God in the world." The world's pain is God's pain, and we are "challenged to participate in the sufferings of God at the hands of a godless world."

Author Philip Yancey tells the story of a young woman he knew called Martha who had been diagnosed with an incurable muscular degenerative illness which made her completely dependent on others. Yancey admits that to talk of the Christian hope of "eternal life, ultimate healing and resurrection" in such circumstances sounded "hollow and frail." Martha could not reconcile what was happening to her with a loving God.

Then came the turnaround. A Christian fellowship group of 16 women, working in teams, committed themselves to her total care. They "listened to her ravings and complaints ... sat beside her all night to listen to her breathing, prayed for her, and loved her. They were available to calm her fears. They gave her a sense of place so that she no longer felt helpless, and gave meaning to her suffering."

But even more significantly than this, Yancey says, is that to Martha "they were God's body." These women were the Christian hope made present and real, "the love of God enfleshed in his body." As a result, Yancey says, she decided to become a Christian in a moving baptismal service just before she died.

Reflecting on the events of 11 September 2001 in answer to the question "Where was God on that day?" then Archbishop of Canterbury Rowan Williams said that people wanted an answer to the question "Where was God visible?" For him, like Yancey, part of the answer lies in the way people are able to become "the love of God enfleshed in his body":

> Where can we see God? ... in part we can see him in those last messages sent to family from people preparing for their deaths ... in

the matter-of-fact human heroism of the people who were involved in rescue work ... in the very simple acts of support and kindness which people offered to other people in the middle of it all.

And surely we have seen God at work in the great outpouring of compassion and support for those affected by natural disasters. But what of those who are left to grieve when their loved ones are taken from them? How are they to understand the Christian hope in the face of such tragic loss?

For psychologist and priest Dr Barry Rogers, who lost his wife to cancer, his main reference point in coping with the loss was the incarnation. "Jesus weeps, gets angry, is moved to act as Lord of life when death disfigures friends and distorts personal relationships." But Jesus' own death is not the end of the story. God raised him to new life, finally overcoming the sting of death for all humanity. "His resurrection underlines the foundational truth that God remembers us, recalling us to mind always; the pattern of self-hood—dissolved (but not destroyed) at death will be recreated elsewhere," Rogers says.

Part of C. S. Lewis' agony in losing his wife to cancer was wondering whether he would ever be reunited with her. Towards the end of *A Grief Observed* he accepts there are limits to what he can know, but his reflection is now more hopeful than when he at first experienced "the double bolted door" and silence.

> When I lay these questions before God I get no answer. But a rather special sort of "No answer." It is not the locked door. It is more like a silent, certainly not uncompassionate, gaze. As though he shook his head not in refusal but waiving the question. Like, "Peace, child; you don't understand."

This is perhaps the peace which passes all understanding—until the parousia. Perhaps we, like Job, until then can only confess to uttering "what I did not understand ... which I did not know" (Job 42:3).

Future Challenges:
Technology, Robots, and Caring for Creation

18. Can the computer be a substitute for humanity?

By Mark Worthing

When world chess champion Garry Kasparov was beaten by the IBM computer Deep Blue in 1996, concerns were raised about our relationship with computers and how human uniqueness could be under threat.

In the early years of computing there was a very real fear among some, expressed in both academic essays and sci-fi writing, that computers might achieve consciousness, and rise up to challenge human beings. Even though films like the *Terminator* and *Matrix* series have been built upon this scenario, few now realistically consider such a revolt of the machines a possibility. The present concern is rather that elements of our humanity will be all too easily allowed to diminish as we happily allow computers to do much of our thinking, remembering and decision making on our behalf.

One theologian is reported to have quipped in the early days of the PC that computers have no more theological significance than typewriters. But are computers simply complex human tools—or are they something more than this? Are the products of the pinnacle of God's creation surpassing that pinnacle themselves?

In just a single generation the advent of computers has produced a host of changes to the way we think, learn and act as modern humans. Those of my own generation and older were accustomed to memorising great slabs of material. At school we memorised everything from classic poems, to the periodic chart, to times tables, nations and their capitals, etc. At Sunday school and church we memorised Bible verses (and not just two or three favourites, the books of the Bible, the Lord's Prayer, the Apostles and Nicene Creeds, and often the contents of one or the other of the classic catechisms). We trained our minds from an early age to store great amounts of material and to recall this information when needed. Our children and grandchildren are neither taught to memorise information nor do they see the need to do so. That's what the internet is for, they tell us. They can find anything they wish, whenever they wish. And if asked to say the Lord's Prayer or the Nicene Creed at church, someone will put it up on the screen.

Memory is not just about storing and recalling information; it is about how we relate to that information and how that information shapes us.

Future Challenges: Technology, Robots, and Caring for Creation

When computers began to allow us to rely on them to store our information, it meant that there would be changes in our human relationship to this information. Yet this and other changes to our way of thinking, acting and living in the computer age have largely not been reflected on theologically.

A recent children's movie called *Wall-e* portrayed the remnants of human society living on a deluxe space ship for several generations with computers and other machines doing everything for them. Apart from being too heavy and too weak to get about the ship unaided, these humans had also become weak intellectually and creatively as these functions were all happily turned over to computers and other machines. We have been happy to let computers do our complex (and even not so complex) maths for us, to store our data, organise our days, weeks and years, reminding us when we need to go to an appointment or finish a project; we have let them store the verses of Scripture previous generations would have learned by heart, as well as our most significant poems and stories. But are we happy with this? What other functions might we be willing to hand over? Decision-making, political views, creativity ... faith?

The perceived threat to our human uniqueness was illustrated by the famous encounter between then world chess champion Garry Kasparov and the IBM computer Deep Blue in a match in February 1996. Before the match, Kasparov stated that he saw himself as the defender of humanity and human ability against the onslaught of computers. The kind of thinking required by chess, after all, is so uniquely human that even the best of computers would only be a pale and predictable imitation of the best human play—or so it was thought. The first game with Deep Blue was a wake-up call. Deep Blue won. For the first time a computer, albeit one capable of processing 50 billion possible moves within three minutes, had beaten a reigning world chess champion at a long play game of classical chess.

Analytical thinking has for centuries been one of the chief attributes of philosophical and theological anthropology and is said to distinguish humans from the rest of the animal kingdom. Yet when a computer can beat the best that humanity can produce at our most symbolic test of analytical thought, where does that leave humans and our perceived uniqueness? At the very least, the advent of something approaching artificial intelligence has the potential to engender an identity crisis for modern humanity.

For students of artificial intelligence or AI, the Kasparov versus Deep Blue game was a hugely significant incident. Many immediately thought of Alan Turing and his Turing Test for Artificial Intelligence. Alan Turing,

Can the computer be a substitute for humanity?

famous for helping crack the enigma code in WWII, posited that when a human being, asking questions of a computer and a human subject veiled behind a screen, could not distinguish which one was human, then AI had been achieved.

If AI is possible, if the best computer can beat the best human player at chess, do we need to rethink our human uniqueness? It is hard enough when we feel our jobs can be replaced by computers, but is our human identity also under threat? This was a big question in the '60s and '70s in the early days of computing when imaginations ran wild about what was possible. Ironically perhaps, now that computers have exceeded many of our early expectations, we as a species seem much less concerned about the uniqueness of our human identity being threatened. We are more comfortable with the technology, and have learned that our personal computers, unlike HAL in *2001: A Space Odyssey*, are not going to gain sentience and try to shut us down before we shut them down.

From a theological perspective, we have perhaps become more confident that our true uniqueness lies in our relationship with God, not in our ability to play chess, store large amounts of data, spell-check, control traffic patterns, or run programs. But we should not be surprised that human beings have been able to create something that can perform many functions we once believed only people could do. The theologian Philip Hefner coined the term "created co-creators" some decades ago to describe our ability and propensity to reflect the image of God by ourselves creating marvellous things—including perhaps even AI. This does not make us God any more than it makes computers genuinely sentient—but it does say much about human beings made in God's image. We should neither fear for, nor worship our own creations, but rather be in worshipful awe in the presence of the one who created beings capable of reflecting God's own creativity so powerfully.

The onslaught of computers in our world and our daily lives does not signal the need for a headlong retreat from technology, nor a head-in-the-sand approach. The Christian community needs a genuine, working theology of the computer to help us understand just what it is we have created, and what it means for our world, our view of God, our own humanness and the future of humanity.

Future Challenges: Technology, Robots, and Caring for Creation

This chapter is an edited extract from "Computers, God and Humanity—Toward a Theology of the Computer," the 2012 ISCAST Victoria public lecture on Science and Faith, delivered by Mark Worthing on 9 November 2012 at the Centre for Theology and Ministry, Melbourne, Australia

19. Do robots spell the end for the human race?

By Daniel O'Leary

Could artificial intelligence (AI) and the development of robots that can think and feel threaten the survival of the human race, as scientist Stephen Hawking has warned, or could they be the work of the Holy Spirit?

On 9 March 2016 AlphaGo, a computer program designed by the British artificial intelligence company DeepMind, beat Lee Sedol, the world's top-rated AlphaGo player. AlphaGo is an ancient chess-type Chinese board game.

Why was this event so widely reported in social media? Because, as the *Times* leader said:

The computer was not programmed how to play the game; it taught itself … AlphaGo can act, but it can also react. It can use intuition and anticipate the possible long-term effects of its action. In short, it can think.

Also in March 2016, a socially intelligent human-like robot, Nadine, was unveiled by scientists at Nanyang Technological University in Singapore. According to a news release at the time, Nadine has "her own personality, moods and emotions, soft skin and brunette hair, good memory, smiling eyes when greeting you and shaking hands … like a real companion who is always with you." Because of Japan's 35 million elderly people in care facilities, the government is leading the AI research field, pouring millions into elderly-care robotics development.

Welcome to the cyborg society. It is claimed that robots such as AlphaGo and Nadine will transform everyday life for millions by 2025, and will match human intelligence by 2045.

Such speculation about AI is spreading excitement, wonder and anxiety on a daily basis. Two recent films offering startling insights into human/machine relationships and possibilities—*Her* and *Ex Machina*—have captured the popular imagination.

Bill Gates, co-founder of Microsoft, believes that "robotics is the next big thing." It will include "a whole lot of complex apps, central to our lives, that we cannot even conceive of now," according to Peter Donnelly, professor of statistical science at the University of Oxford. A remote-controlled Rolling

Future Challenges: Technology, Robots, and Caring for Creation

Bot that acts as home security guard and child-minder has been unveiled by LG Electronics and was seen at the Mobile World Congress trade fair in Barcelona.

Hod Lipson, professor of mechanical engineering at Columbia University, is exploring ways of developing a robot that can design itself, learning in the same way a child does and gradually evolving like a species. His team is developing self-aware robots—machines that can "figure out how to walk, develop a sense of what they look like, and even learn to self-replicate." He expects machines to emulate empathy and develop common sense, eventually outstripping human intelligence. UK scientist Brian Cox is upbeat and positive about the current evolution of such expectations.

Many anxious observers wish to turn the spotlight on the threat posed by such a future. The dystopian forecast of many experts points to huge unemployment, to serious social unrest and, eventually, to humanity's destruction. Wendell Wallach of Yale University's Interdisciplinary Center for Bioethics has warned that the technological advances have now made killer robots a possibility in that they could initiate lethal activity. Scientist Stephen Hawking warned the world in 2014 that AI is a threat to human existence. "The development of full artificial intelligence," he said, "could spell the end of the human race."

But can that day ever come when artificial intelligence will equal or transcend emotional intelligence, spiritual intelligence and human imagination? Can robots be built that share our evolutionary biology? Most Christian research fellows believe that you can simulate aspects of the human experience in machines but you cannot actually recreate human subjectivity. From the moment you arrive in the world to the moment you die, you're learning and feeling, while evolving through the relationships around you. To imagine such a multiplicity being fully simulated in an artificial entity, they hold, is a serious misunderstanding of what it means to be human.

In the meantime, technological giants such as Google, Microsoft and Apple are investing billions in AI research. The focus is on practical possibilities and huge profits.

But the deeper philosophical and theological issues are only now beginning to be debated in earnest. In the life of the world, for instance, could this human/machine interaction constitute a most significant breakthrough in evolutionary progression? Given the astonishing implications of

Do robots spell the end for the human race?

the incarnation, could AI be providential—a unifying, spiritual threshold in the unfolding spirit of the human quest?

A BBC Radio 4 program, *Beyond Belief*, featured many such questions: Will advanced, supremely intelligent robots become self-aware, enjoy relationships—or simply simulate these human states? Will AI enable a sense of compassion, justice and wonder? If humanity is created in "God's image," will robots share that distinction too, their very invention seen as the work of the Holy Spirit incarnate? Will they one day enjoy free will, capable of opting for evil as well as good? Or is it still silly to ask such questions?

At a recent retreat, a participant was describing a brief, profound and intense moment. "I knew it was a precious experience of love," she said. "I had been present at the birth of my young friend's son, Ala. He was lying on a mat, and I was looking into his eyes. I became aware that we were not just looking into each other's eyes but into the depths of each other, and really communicating for a timeless moment." What would his nanny-robot make of that moment?

It is Sunday morning. The church is full of human beings, some in seventh heaven because they have had wonderful experiences of falling in love, passing an exam, holding a new baby, being deeply moved by a beautiful poem or film; others carry a sadness, a loss, a failure, a broken dream, a new addiction. As he distributes the bread and wine with great carefulness, will the robot parish priest have enough "soul" to resonate with the infinite complexities of the human beings lined up before him?

Pope Francis has said that we are not entering an era of change, but a change of era. The recent ripples of gravity detected from black holes have led scientists to herald a new age for astronomy. The expectation of finding intelligent life on distant planets may be the dawning of a radical transformation in human self-understanding. And the swiftly evolving expectations around AI are leading experts to predict another astonishing, though ambiguous paradigm shift for the human race.

> This chapter was first published as an article in *The Tablet* (The International Catholic Weekly) of 9 April 2016, with the title "Stranger than fiction" (see www.thetablet.co.UK) and is reproduced with permission of the publisher.
> Fr Daniel O'Leary's website is www.djoleary.com

20. God is not just the God of human beings

By John Pilbrow

Theology can often suggest that God is only to be found in human beings and human encounters, but what about the rest of creation? This chapter reports on several keynote speakers at a 2015 conference* which explored this question.

The conference theme, Rediscovering the Spiritual in God's Creation, was deliberately chosen because it often seems that something important is missing from our thinking about God's creation. Retired American pastor and theologian, Paul Santmire, explained that this results from the spirit/matter dualism of classical Christian theology originating with Augustine. That is, "things spiritual are good and things material are of lesser value, or even of no value." Such a view devalues the material "world of nature, in general, and the human body, in particular" and implies that God is radically separated from his creation. However, in agreement with Luther, the triune God is seen as "sublimely natural" and "present to the whole cosmos." Santmire reminded us that while our concept of God must mirror the ever-expanding cosmos, it is often quite limited and geocentric. He says we need reminding that "God is not just the God of human encounters."

Denis Edwards (Australian Catholic University), reflected on the sense of awe he experiences walking in the Flinders Ranges, and he asked whether our encounters with birds, trees, forests, deserts and beaches can be considered true encounters with the triune God. With Athanasius, he sees God as both transcendent and "radically immanent to all things through the word (or wisdom) of God (Jn 1:3)," enabling "creatures to be, to inter-relate, and to become." Edwards argued that without revelation we cannot properly understand mystery and transcendence when encountering other creatures in our world. Though characterised by competition and ambiguity, the world is nevertheless beautiful.

Edwards' emphasis on seeing God's presence in the natural world was echoed by Heather Eaton (Saint Paul University, Ottawa). Since, in her view, modern theology has not caught up with modern science and our liturgies do not involve serious engagement with the sciences, she argued "for a renewed Christian vision and a larger horizon for Christian theology."

God is not just the God of human beings

Therefore, dialogue between science and religious insights is urgently needed.

Ernst Conradie (University of the Western Cape, South Africa) sought an enriched liturgy that acknowledges God's "counter-intuitive presence" in creation, while still emphasising the centrality of the cross of Jesus Christ. This vision has "far-reaching implications for economic inequalities and ecological destruction." He spoke of an "emerging horizon" that "enables one to see further from within a specific liturgical location, to surmise what lies beyond the horizon, where land and sky, earth and heaven meet," and enables us to see God and the world from God's perspective.

Taking her cue from Colossians 2:21, "Do not handle, do not taste, do not touch," Vicki Balabanski (Flinders University) asserted that we must acknowledge "our interconnectedness with our biology, our humanity and the earth as a whole." She explained that we all host a miniature cosmos of about 100 trillion microbes, some of which assist in digestion of food while others defend us against pathogens and are essential to our health and well-being. Reminding us that the early Christians had to rethink issues of purity "in an interconnected world whose creator had declared everything good," our theology today must recognise that "all things, visible and invisible to the human eye, are recognised as being created and sustained in Christ (Col 1:15–20)."

Focussing particularly on Job 28, Norman Habel (Flinders University) asked what we can learn from the "ancient wisdom school, the 'scientists' of their day," who were committed to discovering the social and natural dimensions of their world. Taking a cue from the narrator in Job 28, who sought to understand the nature of innate wisdom and its locus in creation, he argued that our search for the "spiritual" dimension of wisdom concerning the natural world must be sought through the "scientific" method of observation.

Celia Deane-Drummond (Notre Dame University, USA) turned our attention to the theological significance of research concerning the "close evolutionary and ecological entanglement between humans and other social species." She placed this in a context where the work of the Spirit in creation is seen as an on-going "theo-drama" in which the wisdom of creation and the wisdom of the cross are best understood in the light of redemption. The specific example she chose concerned recent studies of hyena-human interactions that took the conversation well beyond familiar domesticated animals. She claimed that evidence of "wild justice" or "altruism" in animal behaviours invites "reappraisal of the theological significance of creatures more generally" and impinges on our "divine mandate to bear the image of God." Deane-Drummond also joined the chorus seeking better engagement between theology and contemporary science.

In considering "so-called" natural disasters, Cambridge geophysicist and a world authority on volcanoes, Bob White (Faraday Institute and Department of Earth Sciences, Cambridge University) reminded us that while earthquakes, floods and volcanic eruptions contribute to making our world fertile, the thousands of deaths that sometimes occur have usually been exacerbated by human actions. Such events certainly "bring into sharp focus the relationship between the creator, his creation and humans made 'in his image.'" While God rules over creation, White insists it is inescapable that we must continue to work for "a better understanding of disasters, to enable communities to build resilience against them, and to strive to remove unjust disparities in wealth and resources that cause the poor to suffer most."

The conference concluded with the adoption of the Serafino Declaration (see box below) which outlined the major spiritual dimensions of God's creation recognised during the course of the conference. The natural world is to be treasured for what it is, rather than just for what it does. It has integrity and is therefore to be respected and cared for precisely because it is God's creation.

- * The conference was run by the Graeme Clark Research Institute, Tabor College, Adelaide, and held at the Serafino Winery, McLaren Vale, 10–13 March 2015. (In association with the Faraday Institute, Cambridge, The Charles Strong Memorial Trust, Catholic Earthcare Australia, The Australian Catholic University and Catholic Education SA.)

An extract from The Serafino Declaration

We were challenged to consider a range of the diverse core insights and calls to action:

- The Healing of Earth: to revive our consciousness of the sacredness of our planet and make the healing of our environment a primary mission.
- The Narrative of Landscape: to recognise the precedent of Aboriginal peoples as custodians of country with an acute sense of responsibility for maintaining its integrity.

God is not just the God of human beings

- The Spirituality of Country: to acknowledge that the churches have undervalued this spiritual consciousness and should now work towards reconciliation.
- God's Presence in Creation: to celebrate and experience God's presence at the heart of creation.
- Christ in Creation: to discern the bodily presence of Christ in all communities of nature, from the nano-cosmos of our bodies to the macro-cosmos of the universe.
- The Spirit in Creation: to experience the Spirit guiding us to relate to fellow creatures as participants with us in a cosmic theo-drama grounded in places to which we belong.
- Wisdom in Creation: to experience innate wisdom as a cosmic spiritual reality, as Job once did (Job 42:2–5), by exploring the domains, the laws and forces of nature.
- Disasters in Nature: to prepare for such natural disasters by enabling communities to build resilience against them and by removing unjust disparities in the use of Earth's resources.
- The Rights of Nature: to respect all domains of nature and recognise their intrinsic rights as valued components of the cosmos, whether they be galaxies or gardens, coral reefs or rainforests.
- The Consciousness of Earth Beings: to accept our identity as Earth beings and to celebrate Earth as our source and our home within an interconnected community of life and kin.
- An Emerging Horizon of Hope: to reconsider ways of looking at the world around us that legitimise current constellations of power and to express the hope that a different world is possible.

See the full declaration at: http://seminaryalliance.org/wp-content/uploads/the-Serafino-Declaration.pdf

21. God's sustaining power at the heart of life

By Donald MacKay

Just as an image on a TV screen is maintained by a sustained pattern of regular signals, so our existence hangs moment by moment on the continuance of the upholding power of God.

"I believe in one God, maker of heaven and earth." So says the ancient creed of the Christian church. The claim of biblical theism is that the world in which we find ourselves is not eternally self-sufficient: it has a maker, on whom it depends not just for some initial impulse long ago, but for its daily continuance now.

This is strange language to modern ears. The world we know seems very stable, reasonably law-abiding (in the non-human domain at least) and not at all obviously in need of any divine power to keep it going. Over the past 200 years and more, we have become accustomed to thinking of it as a mechanism, intricate perhaps beyond the grasp of human understanding, but still something self-running and self-contained.

Thinking in these terms, we might see some point in bringing in God as the original creator of the universe; but we might find it particularly hard to visualise any sense in which a universe, once created, could continue to depend on its creator for its existence.

Without pretending to fathom the mysterious depths of these biblical claims, I believe we can get some feeling for their meaning from the imagery of modern physics. Ask a physicist to describe what he finds as he probes deeper and deeper into the fine structure of our solid world, and he will tell you a story of an increasingly dynamic character.

Instead of a frozen stillness, he discovers a buzz of activity that seems to intensify with increasing magnification. The molecules he pictures as the stuff of the chair you are sitting on—and of the body sitting in it—are all believed to be in violent motion, vibrating millions of times in a second, or even careering about in apparent disarray, with an energy depending on the temperature.

Each of the atoms composing those molecules is thought of as a theatre of even more dramatic activity, likened by Niels Bohr to the whirling of tiny planets around a central sun, but nowadays pictured as the vibrations of a

God's sustaining power at the heart of life

cloud whose shape and density determine the probability of various kinds of discrete events called light emission, electron absorption and the like.

Modern physics says it is to such elementary events—myriads of them, continually recurring—that we owe all our experience of the solid world of objects. Even the fundamental particles postulated by theoretical physicists as the building-bricks of our world are thought of as spending their time in snapping from one to another of a variety of different states, or even in continually exchanging identities.

For our present purpose, it does not matter for how long physics is likely to go on using these particular images. Their relevance here is merely to illustrate a key concept that, I think, may help us to grasp what the biblical writers mean when they say that the stable existence of our world depends on the creative activity of God. We can call it dynamic stability.

In our everyday experience, chairs, tables, and rocks are typically stable objects. They are there. Nothing may seem to be happening to them or in them for most of their existence; yet the modern physicist is quite content to describe such stable objects as a concurrence of unimaginably complex and dramatic sub-microscopic events, without any suggestion that he is contradicting the facts of experience.

All he claims is that their stability is not static but dynamic. The quiet solidity of physical objects, he would say, reflects the coherence of uncountable myriads of events at the atomic or subatomic level, each of which, by itself, might seem almost unrelated to its neighbours in space or time.

For another and rather different illustration of dynamic stability, ask a television engineer to explain the patterns of light and shade that form the image on the face of a TV set. All that is happening on the screen, he will assure us, is but a succession of isolated sparks of light produced by electron impact; yet, because of the regularities in the program of signals controlling the intensity of the beam of electrons, these sparks fall into a coherent pattern, forming stable images of the objects we are watching, whether the scene is one of violent change or of perfect calm—or, indeed, whether it continues in being at all; all depends entirely on the modulating program. Any stability the picture has is a dynamic or contingent stability, conditional on the maintenance in being, and the coherence of the succession of event-giving signals.

I need hardly say that none of these examples of dynamic stability is meant as an explanatory model of our mysterious dependence on God as portrayed in the Bible. But if we ask the writers of the Bible what makes our world tick—the sort of question that underlies any attempt to build a science of nature—we will find them using remarkably similar language.

From the biblical standpoint, all the contents of our world, ourselves included, have to be "held in being" by the continual exercise of God's sustaining power. In Christ, says Paul, "all things were created, in heaven and on earth, visible and invisible … all things were created through him and for him. He is before all things, and in him all things hold together" (Col 1:16–17).

Or, as the writer to the Hebrews puts it:

In these last days [God] has spoken to us by a Son, whom he appointed the heir of all things, through whom also he created the world. He reflects the glory of God and bears the very stamp of his nature, upholding the universe by his word of power. (Heb 1:2–3)

For biblical theism, then, it is clear that the continuing existence of our world is not something to be taken for granted. Rather, it hangs moment by moment on the continuance of the upholding word of power of its creator, as dependent on this as the picture on a TV screen is on the maintaining program of signals.

This chapter is an edited extract of *Science, Chance, and Providence: The Riddell Memorial Lectures*, forty-sixth series delivered at the University of Newcastle upon Tyne on 15, 16 and 17 March 1977. NUA/10/2/4, Newcastle University Archives, Newcastle University Library, UK.

22. Climate science and Christian faith share much in common

By Chris Mulherin

What is the place of scepticism and belief in climate science and in Christian faith?

The battle between sceptics and believers is hotting up. One side says it's all hogwash—a story foisted on us by credulous people who need certainty in their lives and a cause to believe in. Our responsibility, they say, is to ignore the doomsayers and get on with our lives. The other side says that while there is no ultimate proof, belief arises logically as the best explanation of the evidence. They say that in the absence of a better theory, we are morally obliged to heed the precautionary principle and commit ourselves to the cause.

As a discerning reader you will realise that I'm playing with ambiguity in such talk of scepticism, belief and morality. Are we talking of climate science or religious faith?

Belief in humanly caused climate change shares many characteristics with religious belief, a similarity that makes a lie of the claim that science and faithful belief are incompatible. And it is incorrect to suppose that science deals only in proofs and certainties and is a source of incontestable knowledge; the history of science is littered with "scientific" beliefs once held to be true and now considered to be false.

The climate wars

If you aren't up to speed on the climate debate, the gist is this: certain gases in the atmosphere act as an insulating blanket around the earth; they trap heat in a similar manner to the glass walls of a greenhouse. Life on earth depends on this "greenhouse effect"; without our terrestrial blanket the earth would be a perpetually uninhabitable glacial desert. So far so good—at least for the millennia prior to the industrial revolution. But human industrial capacity produces greenhouse gases, particularly carbon dioxide and methane, which contribute to the effectiveness of the insulating blanket.

So the fundamental question at the heart of the climate debate is this: how significant is the human contribution to global warming? This is the key question tackled by the Intergovernmental Panel on Climate Change

(IPCC), which, since 1988, has collated the research of thousands of scientists under the auspices of the United Nations.

Climate sceptics (or contrarians or, in the extreme, deniers) doubt that human activities have a significant influence on global climate. On the other hand, believers accept the opinion of the IPCC that it is probable we humans are contributing significantly to a dangerous increase in global temperatures. The stakes are enormous and we proceed with business as usual at our peril; while the evidence does not amount to certain proof, it is beyond reasonable doubt and leaves no room for delay.

Climate science and Christian faith

Let's turn now to connections between climate science and Christian faith. On the one hand, it is clear that theology and science are not examining the same objects; they pursue truth in different realms and they use different methods to arrive at their conclusions. But notwithstanding these differences, there are many similarities in the way science and theology go about their business.

In the first place, *both climate science and Christianity claim to be true about the things of which they speak.* While doubts or lack of certainty remain, both science and Christian faith make claims that are either true or false. It is either true or false that humans are causing significant changes to the global climate; it is either true or false that Jesus Christ was God incarnate and participated in the creation of the universe. But neither of these claims can be *proven* to be true or false; both are beliefs held with varying degrees of confidence.

Secondly, *both scientific and Christian beliefs are based on evidence—* neither consists of irrational or blind belief. Science, for the most part, focuses on the empirical evidence of the senses. It weighs up evidence using a number of foundational or "pre-scientific" assumptions about the order and rationality of the universe, as well as the human ability to know that universe. And wherever possible science turns to experiment to test its conclusions. On the other hand, Christianity, like historical or moral reasoning for example, does not rely on repeatable experiments in order to make its case. But, just as history and philosophy have their own traditions and norms of enquiry, so too theology is a rigorous discipline.

Part of its rigour arises from the fact that *Christian thought, like climate science, is a collaborative effort* based on networks of mutual trust. No one person can master any serious field of enquiry. Thousands of scientists contribute to the climate discussion, each one an expert in their own field. But they have to trust the judgements of others with whom they collaborate in order to contribute to the overall IPCC "opinion." So too, no theologian or biblical scholar or philosopher of religion can hope to master any but their own small corner of "the knowledge of God."

Within both areas of enquiry there is also *room for dissent and a healthy questioning* that challenges the accepted norms in a way that strengthens the edifice of belief—either by confirming it or by showing where some beliefs are found wanting and ought to be rejected. It is true that there is more diversity of belief concerning "the God question," but that is not surprising given that the moral and personal stakes are higher, the evidence is less tangible, and the subject matter is further from the everyday experience of the senses. And whether it's science or theology, not only can the evidence be interpreted in different ways (which it often is), but even working out what constitutes significant evidence as opposed to irrelevant "noise" is an interpretive judgement.

Finally, this room for dissent means that *neither set of beliefs is capable of offering "final proofs" to the committed sceptic.* Renowned scientist, Michael Polanyi, describes scientific belief in a way equally appropriate to religious belief: it is about achieving "a frame of mind in which I may hold firmly to what I believe to be true, even though I know that it might conceivably be false." This lack of ultimate proof is found in the reports of the IPCC, which speak of degrees of likelihood, reflecting the fact that any scientific statement is, in the final analysis, the considered judgement of fallible human beings. In the words of physicist Richard Feynman, "scientific knowledge is a body of statements of varying degrees of certainty—some most unsure, some nearly sure, but none absolutely certain."

The climate sceptic, along with many who are sceptical about religious belief, assumes that reasons must be so overwhelming that the believer is left no possibility for doubt. But climate science, like Christianity, can always be questioned further and no amount of evidence will convince the true non-believer. The end to unbelief only comes when my considered opinion is that I have heard enough and that a conclusion is warranted— when I choose to believe what I know could conceivably be false.

23. Why a Christian response to climate change?

By Chris Mulherin

Christianity, not science, provides the moral
basis for action on climate change.

It had been a good semester in Introduction to Climate Change. The students attended class, they submitted assignments, and—icing on the cake—they chatted enthusiastically about greenhouse gases, solar irradiance, black body radiation and ocean acidification. Most were convinced that humans were causing dangerous levels of global warming and that something must be done.

Enter Rick—week 11—in a class discussion about the future.

"I understand about the science and that humans are probably causing the problem. But why should we do anything about it?"

There is a lengthy silence—some students are surprised at such foolishness. For others it dawns on them that Rick is onto something.

In the previous chapter I drew parallels between climate change and Christian faith, suggesting that they were both examples where certainty is impossible and yet the issues in question are life-changing. In the case of the climate, scientists are confident that humans are warming the planet, while sceptics demand unreasonable levels of certainty.

Let's turn to Rick's not-so-innocuous question about the moral reasons for action. On what grounds would former Prime Minister of Australia, Kevin Rudd, famously refer to climate change as "the great moral challenge of our generation"?

A high stakes game

There is no doubt that the global warming stakes are high. According to *The Guardian*, climate change is already responsible, every year, for 400,000 deaths and over US$1.2 trillion in costs. That's the situation now, with global temperatures only 0.7 degrees Celsius above pre-industrial levels.

A global temperature rise of 2 degrees Celsius by the end of the century seems inevitable. A rise of 4 degrees would be catastrophic, leading to widespread human and ecological devastation. And the consequences would fall disproportionately on those in the less developed world as they pay the price for 200 years of growth of the now-rich nations. This is an

Why a Christian response to climate change?

"ecological debt" owed to those whose resources have been obtained at bargain prices to fuel the health and wealth of energy-hungry industrialised societies.

Climate change forces us to recognise that how we live today deeply affects the people of the planet who are distant from us in space and time. And it tells us what we ought to have known long ago; unrestrained economic growth, dependent on resource use, is unsustainable.

The climate debate particularly matters for Australia because we have the highest emissions per capita of any country in the developed world. We are the 15th largest emitter and, if our coal exports are included, we are responsible for almost 5 percent of global carbon emissions. All this in the country that has the highest median adult wealth in the world.

Solutions are hard to come by

Not only are the consequences of global warming grave, but solutions involve difficult commitments. While science offers scenarios, it cannot make the decisions for us. And a confidence in market forces would be naïve; the market has no interest in time scales of 50 to 100 years, or of fundamentally changing the relationship that humanity has with the natural environment. For the first time ever, if the majority of humanity does not agree to act together, we risk global catastrophe. For the first time in the history of the market economy we face the fact that we have to renounce our aspiration to grow ever richer and more comfortable. And, while some progress is being made, most recently evidenced by the Paris agreement, the task is not acknowledged by all; the current us president is a climate sceptic who has appointed someone of like mind to run the us climate agency.

Now put yourself in student Rick's shoes. Let's say Rick is a product of the secular age, a man of no fixed religion with a basic commitment to his own welfare above all else. He is also a man of science who has imbibed of naturalistic wells. He knows that nature is red in tooth and claw; the lesson of evolution is that the fit survive. Rick also knows that while climate change might be a concern on a global and generational scale, it will hardly touch his own comfortable life in the Antipodes. A dose of retail therapy to buy another air-conditioner will take his mind off the problem.

So, while much current debate focuses on the sceptic's challenge to climate science, hiding in the wings lurks the moral sceptic. Rick has no moral convictions robust enough to induce a change in lifestyle or voting preference. He simply has no serious reasons for acting on behalf of future generations of animals (including humans), or on behalf of a planet that will one day disappear into the cosmic nothingness out of which it was spawned.

In the absence of deep moral convictions we are also faced with what is known as the *tragedy of the commons*, a reference to public land on which cattle and sheep might graze. This common resource becomes depleted because, while it is in *everyone's* long-term interest to care for it, in the short-term it is in *no one's* self-interest to do so. On a global scale there are over 200 sovereign territories, all competing for growth. While it is in their collective interest to curb climate change this century, it is in none of their interests to do so in the next decade or election cycle.

The crucial question then is not *whether* human-caused climate change is occurring; it is not *how* climate change is occurring; it is not what solutions are open to us. The crucial question, presupposed by all discussion, is the one Rick asked the class: "*Why should* we do anything about climate change?"

The is-ought problem

Unbeknown to Rick, he puts his finger squarely on a philosophical problem highlighted 350 years ago by a Scottish polymath. David Hume railed against "vulgar systems of morality" that would magically derive an "ought" from an "is." In simple terms, no amount of science and economics, no amount of discussion about grave consequences, no facts about lost species or deaths attributable to global warming; none of these can tell us what we *ought* to do about it.

In a broadcast of ABC TV's *Q&A*, (February 2013), physicist and "New Atheist," Lawrence Krauss, was asked about the relationship between science and ethics. He was quick to respond that science is highly moral and offers us a "better" world. But, in his concern to paint science as superior to religion and philosophy, he neglected to say that science of itself cannot tell us the meaning of "better." Rick's "better" world, based on a survival of the fittest morality, might be one in which billions of humans perish, so easing the ecological congestion we are currently suffering.

To quote biologist Richard Dawkins, the guru of the "New Atheism':

> In a universe of blind physical forces and genetic replication, some people are going to get hurt, other people are going to get lucky, and you won't find any rhyme or reason in it, nor any justice. The universe we observe has precisely the properties we should expect if there is,

at bottom, no design, no purpose, no evil and no good, nothing but blind, pitiless indifference.

Meanwhile "Morality and Climate Change" was the topic for a conversation at the University of Melbourne (in February 2013) that included Australia's most famous and controversial philosopher Peter Singer. But, again, while our individual moral responsibility was assumed, no reasons were given that might answer Rick's question. When asked at the end of the lecture how people could be motivated to act, Singer replied gloomily that he wished he knew.

The Western clash of worldviews

In the end we face a clash of worldviews. On the one hand is a non-religious evolutionary naturalism that, although it speaks moral language, has no transcendent realities and has no intrinsically moral foundation for ecological action; the ultimate reality is the law of the jungle and the meaningless cosmos.

On the other hand are religious views, particularly the Judeo-Christian perspective that underlies Western moral sensibilities, with its robust foundation rooted in the intrinsic goodness of creation and the inherent dignity of all people. Christian faith overturns common ideas about both human rights and attitudes to the creation, and it offers two fundamental moral reasons for acting on climate change, one rooted in a creation ethic, the other in an ethic of justice.

The earth is the Lord's

"The earth is the Lord's," says the Psalmist, and, for the Christian, the creation is just that—a creation that reveals its author in the wonder and beauty of the natural order. God the creator is sovereign over the cosmos and human creatures have been entrusted with the stewardship of this masterpiece.

Humans are called to live in harmony with the planet and to reject delusions of growth at no cost. We have lived too long with the illusion of mastery that comes with science and the belief that there will always be a technological fix that does not require a change of heart.

In that vein, Thea Ormerod (*The Melbourne Anglican*, February 2013) suggested that increasing extreme weather events are a way the creation has of crying out for "humanity to restrain our consumption in order to protect the ecological balance that allows life to thrive on the planet."

Love your neighbour

The other moral resource that Christians draw on is a view of justice that is not rooted in the useful forensic fiction of human rights, but in the theological truth of the dignity of every human being, each one made in the image of God. Our Lord's call was not to speak the language of rights and to claim what is mine. It was to love our neighbour and especially the one who is poor or blind or the victim of robbery, or the ones dispossessed of their island home due to rising sea levels. It is my planetary neighbours, suffering already and who will do so increasingly, that bear the cost of Rick's air-conditioner or my aeroplane flights.

A Christian response to climate change

Why a specifically Christian response to global warming? Because the Christian faith offers the moral resources that give sound reasons for action. Christianity makes sense of those intuitions we have about social justice, about shared responsibility, about the value of the natural world, about species extinction, about the global good.

It is time for Christians to help our culture recalibrate its moral compass. It is time to admit that materialism and over-consumption and greed are evil. It is time to accept that happiness does not increase with abundant wealth and that he who dies with the most toys does not win, but rather risks losing his soul. It is time to acknowledge that "better" does not mean growing and spending like us. It is time to confess that planet-abuse is an offence to the creator and robbery of our neighbour. And it is time for us in Australia, the world's rich, to remember that our Lord took the side of the poor and the hungry and the dispossessed—those most exposed to the ravages of the groaning creation.

24. Why climate change is real and human-induced

By David Griggs

Climate scientists are pretty much universally of the opinion that climate change is real, that we are seeing a warming of our climate, that the warming we observe is due to human emissions of greenhouse gases such as carbon dioxide caused when we burn fossil fuels, and that we are going to see an acceleration of that warming over coming decades.

So why have we found it so hard to convince everyone else? Well, the most important reason is that the science is complicated, too complicated to explain in a sound-bite or even in an article like this. Our conclusions are not based on one fact or one piece of evidence, but on thousands of lines of enquiry, experiments, observations, models, computations etc.

The bottom line is that since a scientist called John Tyndall, over 150 years ago, first discovered that gases such as carbon dioxide absorb and trap radiation, and the theory of global warming was born, all the evidence has ended up supporting and reinforcing the theory; and no other theory we have been able to come up with has been able to do this. And believe me we have tried—no climate scientist wants global warming to be real, it is not pleasant working day in, day out, being the bearer of bad news. And any scientist who did manage to disprove global warming would gain the kind of fame and fortune usually reserved for film and rock stars and there would be a Nobel Prize to go along with it.

Here are just a few of the pieces of evidence that make us conclude that global warming is real:

- The average temperature at the Earth's surface has continued on an upward trajectory at a rate of 0.17 degrees Celsius per decade over the last three decades;
- Temperatures throughout the lower atmosphere (the troposphere) have also warmed consistent with the surface temperatures;
- The temperature of the top 700 m of the ocean continues to increase;
- Sea level has risen by about 20 cm since the 1880s and at a higher rate over the past two decades (sea level rises due to melting of ice on land and by the fact that water expands when it gets warmer);
- Greenland and West Antarctic ice sheets are losing mass and Arctic sea ice cover shows a clear downward trend;

Future Challenges: Technology, Robots, and Caring for Creation

- Plants and animals are responding to the warming with observed changes in gene pools, species ranges and timing of biological patterns (such as earlier spring flowering).

I could go on and on.

So what are some of the common arguments against human-induced global warming and why don't climate scientists believe them? Again because the science is complicated it is easy to get part of a story or a half truth or to select a piece of data that can make it appear that global warming isn't happening. But these are some of the most common ones:

(1) The world has stopped warming, or variations on this theme. Because of natural variation in the climate the world doesn't warm everywhere at once or every year. But if you look over the long data record or take it decade by decade, which smooths out some of these year-to-year variations, the world has continued to warm.

Temperature measurements measured by two completely independent sources (firstly, thousands of thermometers at the earth's surface over the last 130 years, and, secondly, starting around 1980, annual temperature of the lower atmosphere measured from satellites) shows the same pattern of increasing annual global mean surface temperatures.

(2) The climate is changing or has changed in the past but it's all natural. Climate only changes if something causes it to change. For example, when the sun gets brighter, the planet receives more energy and warms. When volcanoes erupt, they emit particles into the atmosphere which reflect sunlight, and the planet cools for a few years. Natural climate change in the past proves that climate is sensitive to an energy imbalance. Currently, carbon dioxide and other greenhouse gases are imposing a large energy imbalance on the earth system. So, understanding past climate change actually provides evidence for our climate's sensitivity to carbon dioxide and no known natural forcing can explain the recent warming we have seen.

(3) The warming has been caused by the sun or sunspots. Yes, changes in the brightness of the sun causes a change in climate but this effect is calculated to be much smaller than human-induced global warming and this is confirmed by the fact that in the last 35 years of global warming, sun and climate have been going in opposite directions.

(4) Carbon dioxide is only 0.038 percent of the atmosphere. True, but concentrations of carbon dioxide have risen by about 40 percent since the industrial revolution and while most gases in the atmosphere don't trap

Why climate change is real and human-induced

radiation leaving the earth, carbon dioxide and other greenhouse gases do and a small amount has a big effect—that is just physics.

(5) Australia only emits about 1.5 percent of the world's carbon dioxide emissions. Also true, but here lies the crux of the problem. There is no quick fix or magic bullet for climate change; nobody else is going to fix it for us. The only way to stop global warming is to reduce greenhouse gas emissions, which means burning less fossil fuel. Every one of us burns fossil fuel in almost everything we do, from flicking a light switch on to jumping in our car. Alternatively we can absorb greenhouse gases from the atmosphere, by planting trees for example. But because we are all responsible for the emissions we all have to play our part in reducing them. Applying the same argument that we are so small that we can't make a difference, I like to think that my taxes are such a small fraction of the total tax revenue that it can't possibly make a difference if I don't pay it—somehow I don't think the government would agree!

There are many reasons people give for doubting global warming and, when they come up, scientists examine them; so far none has been able to stand up to that examination. The only thing we have left to explain the warming is the greenhouse effect.

So, if we are right, why are we so worried? After all, we are only expecting a few degrees of warming this century. However, it is not the change in the average temperature that will be the problem but the changes in extreme weather and the impacts that would follow. In Australia by the end of the century agricultural productivity would be severely affected, sea levels will rise by up to a metre or more threatening major coastal infrastructure, marine ecosystems such as the Great Barrier Reef will collapse, there will be major loss of biodiversity, we will have more frequent and intense bushfires ... the list goes on and on. As well as the direct environmental impacts there would be compounding economic and social impacts.

Climate change would also have major impacts all over the world, particularly in developing countries, and with the increasingly global economy Australia would not be immune to the fall-out from these impacts as we have seen with the Global Financial Crisis.

Finally, greenhouse gases hang around in the atmosphere for a long time—about 100 years in the case of carbon dioxide. So the longer we delay action the longer we will be faced with the increasingly severe consequences of global warming.

25. It is not only humans that matter to God

By Mick Pope

Scripture supports the view that care for
creation is an act of worship.

When Christians are asked where they go to church, they sometimes say that they worship at St X. Anyone who has been a Christian for any length of time and been involved in a church has probably been through "worship wars"—arguments over what style of music to use in church. In our more lucid moments we are willing to admit that a closer reading of Scripture shows that the whole of life is worship, a service to God (Rom 12:1–2). Having said that, some still think of this worship primarily in terms of evangelism and private piety rather than contribution to larger issues.

We live in a time of ecological crisis, where it is important that we be able to link worship to creation care. The International Union for Conservation of Nature (IUCN) states that the current species extinction rate is some 1000–10,000 times the natural rates at which species disappear. Habitats are being destroyed. Human polluting activities, including greenhouse gases, are destabilising ecosystems, yet biodiversity and healthy ecosystems are what support life and human civilisation.

In the arena of conservation the Christian church has lost ground to "green" thinking, which is generally either agnostic or pantheistic. In their book *The Cross and the Rain Forest* Whelan et al. don't help matters when they adopt a strong dominion model of creation care, where trees are not needed for worship but are simply a source of wood. In accusing Christians who show too much care for creation of being pagan, they are both divisive and ignorant of a deep Scriptural vein of material supporting creation care as an act of worship.

Scriptures prompt us to imagine even the trees as contributing to the worship of God (Isa 44:23) and not just as part of God's earthly temple (Isa 60:13). In that temple, everything that God has created takes its place to testify of his wisdom (Ps 104:23). God loves his diverse creation and its wondrous variety calls forth awe and wonder, "How many are your works?" (v. 24). He cares for it, providing water not only for human agriculture but also for wild creatures outside of the human economic order. In a drying climate due to global warming (to say nothing of natural variability) and the debate over water use in the Murray-Darling basin, it should cause Christians to stop and think carefully not only for pragmatic but also doxological reasons about how water is used.

While Christians often focus on the image of God's appointed gardeners (Gen 2:15), the reminder that there are "wild places" outside of the human economy that God tends directly for their own sake (Ps 104, Job 38–41), should give us a sense of humility that it is not only we humans that matter to God. It should also provide us with some sense of shame and responsibility. In our post-industrial era, human activities dramatically affect the entire planet, and the "groaning" of creation of which Paul speaks, takes on a more urgent meaning (Rom 8:19–20). According to Old Testament scholar Derek Kidner in his commentary on Genesis, a creation without humanity playing its proper role is like a choir grinding on in discord. This speaks against both atheistic and pantheistic claims that humans are merely a cosmic accident, if not one that nature would be better off without. The biblical picture is one of humanity needing to take its rightful role in a redeemed creation, not seeking to escape from it or our responsibility to care for it. If God will liberate creation from its bondage in the future, why not work now for its preservation? Have you ever heard of a Christian recommending moral laxness because our complete sanctification lies in the future?

Finally, the imperative to love our neighbour as ourselves has never been sharper than in a globalised economy where industrialised pollution and environmental damage have extended to the developing world. Greenhouse gases know no national boundaries. Of course, complex problems abound. If we reduce greenhouse gases (and hence impacts on the poor) by eating local foods, we deny growers in developing nations income. Yet consuming cash crops exported to pay unserviceable interest on debts ruins local ecosystems. The link between *eco*systems and *eco*nomics is a close one in God's *oikos* (household). Doing justice means looking after all of God's household so that all people may live in peace and all of creation may flourish in anticipation of God's final redemption.

26. John Houghton: Lessons from a leading climate scientist and Christian

By John Pilbrow

A review of *In The Eye of the Storm*, by Sir John Houghton (2013)

Whether you accept that climate change and global warming are real or remain sceptical, here is a book by a world authority (and professing Christian), that leaves no room for complacency. The author, Sir John Houghton, is an eminent atmospheric scientist who played a major role in putting global warming onto the world's agenda and in helping establish the Intergovernmental Panel on Climate Change (IPCC).

Houghton traces his path from academic scientist (and Oxford professor), whose own research has contributed much to our understanding of global warming and climate change, to the public figure who served as CEO of the British Meteorology Office. The science of global warming and climate change is explained simply for the general reader.

Amongst many honours, Houghton received the 2006 Japan Prize "for contributions to atmospheric science and global warming." In 2007, he accepted the 2007 Nobel Peace Prize on behalf of the IPCC.

Brought up as an evangelical Christian by devout Christian parents, Houghton has remained a committed Christian throughout his life. He believes that science and Christianity complement each other. Thus science is worth pursuing as a God-given activity, and we should welcome the truth wherever we find it. In particular, he sees care for the environment as an inescapable Christian responsibility.

Houghton's research career has encompassed measurements of ozone concentrations in the troposphere (for his doctorate) to ground-breaking measurement of carbon dioxide levels in the atmosphere using miniaturised equipment attached initially to balloons and later to NASA satellites.

Houghton moved into scientific management in 1979 on becoming CEO of the Rutherford Appleton Laboratories, where he developed management and administration skills for raising morale and valuing peoples' expertise. This proved invaluable later as CEO of the British Met Office where he forestalled attempts by those under him—the Sir Humphreys!—to make him merely a figurehead. He chaired the UK Standing Royal Commission on

Environmental Pollution, which reported on public and private transport in 1994.

Gradually it began to dawn on Houghton and others that there was a "growing body of scientific evidence that pointed to human-related carbon dioxide emissions as a major driver of this [climate] change."

Following the establishment of the IPCC, Houghton served as co-chair of its Assessment Panel and lead author of IPCC reports from 1988–2002.

Houghton's research has yielded two specialist texts, *Infrared Physics* (co-authored with S. D. Smith, OUP, 1966) and *Physics of Atmospheres* (3rd edition, CUP, 2009). He also wrote *Global Warming: The Complete Briefing* (3rd edition, CUP, 2004), at a more popular level where a whole chapter is devoted to his reflection on the issues as a Christian.

Houghton has also written two very helpful books at the science–faith interface, *Does God Play Dice?* (IVP, 1988) and *The Search for God: Can Science Help?* (Lion, 1995), the latter based on his 1992 Templeton Lectures at Oxford University.

No stranger to controversy, Houghton has faced opposition from British cabinet ministers, right-wing politicians in the US, and hostility from some fundamentalist Christians, particularly in the southern US states, who deny that humans have had anything to do with climate change! Chapter 16 deals with opposition to IPCC reports, particularly from the oil industry. Much of the invective was directed at Houghton himself.

In 2009, UK journalist Christopher Booker began his book, *The Real Global Warming Disaster* with the words, "unless we announce disasters, no one will listen," attributing the quotation to the 1994 edition of Houghton's, *Global Warming: The Complete Briefing*. This quotation, which is in fact a misquotation, has since been reproduced by many sceptics, including Lord Christopher Monckton and, though not mentioned in *In The Eye of the Storm*, this misquotation also appeared in a piece by Piers Akerman in *The Daily Telegraph* in Sydney back in November 2006. The correct quotation reads,

> If we want a good environmental policy in the future we'll have to have a disaster. It's like safety on public transport. The only way humans will act is if there's been an accident.

Houghton has demonstrated rare wisdom in showing how to function successfully within public and political spheres with integrity. His voice needs to be heard in the churches, particularly where climate change is denied. Christian climate change sceptics need to know that the story of global warming owes much to this unassuming Christian and scientific statesman.

The book ends with an important message for Christians (see the summary below). This is also a book our politicians should read!

Summary challenges

- The world is facing environmental crises of unparalleled magnitude, including some on a global scale.
- Looking after the earth is a God-given responsibility. Not to look after the earth is a sin.
- Christians need to re-emphasise that the doctrines of creation, incarnation, and resurrection belong together. The spiritual is not to be seen as separate from the material. A thoroughgoing theology of the environment needs to be developed.
- Our stewardship of the earth, as Christians, is to be pursued in dependence on and partnership with God.
- The application of science and technology is an important component of stewardship. Humility is an essential ingredient in the pursuit and application of science and technology—and in the exercise of stewardship.
- All of this provides an enormous opportunity for the church, which has too much ignored the earth and the environment and neglected the importance of creation and its place in the overall Christian message.

27. Michael Northcott: Our devotion to idols is killing the planet

By Roland Ashby

Global warming is a major factor in Australia's extreme weather events and is fundamentally a spiritual problem, argues Michael Northcott, Professor of Ethics at Edinburgh University and an Anglican priest. He spoke to Roland Ashby on a visit to Australia in 2011.

Michael Northcott is in no doubt that the Queensland floods of 2010–2011 and cyclone of 2011 have a direct link with climate change. "The main cause is the warming of the ocean to the east of Australia and the resulting increase of water vapour in the atmosphere, which are both the result of global climate change. Cyclone Yasi was almost certainly caused by global warming."

He believes the extent of the flooding was also, in part, "a consequence of unwise clear cutting and deforestation upstate, because trees in a tropical region are flood defences and also act as an air conditioning system. Australia's forestry policy of replacing old growth forests with plantations means that high ground cannot hold the water in extreme storms."

The CSIRO, he says, is predicting that by 2070 Sydney will be six degrees warmer. "So it is going to be very hard to still live there; it will be just too hot. To me this is really ironic because New South Wales is the main source of the coal, which is one of the main drivers of climate change."

Northcott is also concerned about other "climate challenged" parts of the world, including Pakistan and Bangladesh, which have suffered from their own catastrophic flooding, and also parts of North India, which are undergoing extreme climate change. The whole of Sub-Saharan Africa is also finding it difficult to grow enough food for its population. "Droughts which used to occur every 100 years are now occurring every 10 years, and droughts which used to occur every 10 years are now occurring every year."

He is surprised by the number of people in Australia who still dispute the climate change science.

The science is unarguable and 99 percent of climate scientists think it is convincing. Carbon dioxide is a warming gas, and the human race has doubled the amount of carbon dioxide in the atmosphere. What we have done is similar to double glazing a house to keep the heat in. We have effectively put double glazing around the planet. Lord Monckton, Ian Plimer and Rupert Murdoch would have you believe that you can double glaze the planet and it won't make any difference. Unfortunately it's not true.

He also has an answer for those who say that an unusually cold weather spell in the northern hemisphere in 2011 is evidence of global cooling rather than warming. "This colder weather," he says, "was the result of Greenland and the Arctic being too hot. Greenland was six degrees in November and it should have been below zero. This, together with high pressure over the Arctic, pushed the cold air down to north-eastern Europe and the north-eastern United States, and gave them extreme winters."

Well known and well established weather effects, says Northcott, are being ramped up by climate change.

This means we will get more extreme weather, not just warmer weather, but also more extreme cold, as well as at different times of the year, or more wet or more dry. One of the problems is with the phrase "global warming." While average temperatures are rising, there are other places on the planet where it will be colder for a while before it gets hotter, and other places where it will be hotter and maybe cool down a bit. We are just going to get a lot more, of what I sometimes call, "global weirding" rather than global warming. The weather is not going to be consistent and reliable in the way that it once was. It will be weird.

Northcott believes a carbon tax is preferable to carbon trading, or an Emissions Trading Scheme (ETS), and is concerned that the carbon tax announced by the Federal Labor Government in 2010 was proposed simply as a predecessor to carbon trading, which he compares to medieval indulgences. "It's like buying your way out of purgatory without changing your behaviour."

He also describes carbon pricing as "an easy fix" which will have "very unpleasant effects because it will be the poorest people in Australia who pay the biggest price ... if it goes through without a reform to the tax system. Compensation through welfare payments is not good for the poor because it makes them more dependent on the state."

His own preference, he says, is for government regulation of the fossil fuel economy, and Sweden is showing the way:

Sweden has already massively reduced its carbon footprint through government regulation, which is a much faster route to go down. So, Sweden now has a per capita output of about five tonnes of carbon dioxide per person, compared to Australia's 20 plus per person, or the US's 25 tonnes per person. I think the refusal of governments to use the legal controls and instruments available to them is a big part of the problem. In Sweden whole cities are being built with virtually zero-carbon buildings throughout, and electric bicycle and car networks, buses and trains are replacing fossil fuel cars.

At its heart, the ecological crisis is fundamentally a moral and spiritual one, Northcott believes, with the marketing of consumption and the power of advertising a significant factor.

In 1970, Britain's emissions were at a sustainable level. We made most of what we consumed in the country and we didn't consume nearly so many products. And what we did consume, we tended to keep much longer. So people in 1970 might have had a washing machine in their home which was 30 years old and was endlessly repairable by a local mechanic. They might have had a car, which my father did, that they could actually tinker with themselves and fix and keep on the road. They typically commuted to work on trains or buses. There was less than one car per household. Now, what's changed over the last 40 years isn't that people have got happier as a result of having more stuff, it's that they've been persuaded through advertising and the media that having lots more stuff will somehow increase human flourishing, but that hasn't happened.

In theological terms, Western societies are in the grip of idolatry:

In the Old Testament, an idol was something made by human hands that people worshipped and bowed down to, and gave their soul force in the service of. In his book *City of God*, which is an answer to the charge that Christians caused the fall of Rome because they had Christianised the Empire, St Augustine said, "No, what has caused the fall of Rome is Rome's devotion to idols, to luxury wealth and sexual immorality. And furthermore, a people are made a people by the things that they love in common, the common objects of love. It's their devotion to these objects which identifies them, which marks them out as a people who belong together."

Northcott adds that one of our very obvious idols or "common objects of love" today is the car, which is also a major source of daily carbon generation.

If we all significantly reduced our reliance on cars, this, on its own, would give the planet a chance to recover. I like to cycle around Edinburgh as much as possible, but I find cycling in Australia more challenging, because there is so much more pollution here. Cars are not tested here annually for emissions of pollutants; there are no converters, especially on the newest cars, and there are lots of old cars pouring out every conceivable noxious particle.

He also believes that it is a spiritual problem which is at the root of Australia's unwillingness to use its abundant solar resources to tackle the problem.

Here in Australia, you have this fantastic resource called a desert, which gets vast amounts of sunlight. This natural energy could be harnessed and Australia could be the very first completely solar driven country, and yet solar engineers can't get jobs in this country; they end up going overseas for work. I think there's a real spiritual problem in the whole collusion between coal mining, forestry, the car industry and government here. There's a kind of wickedness, and corrupt relationships which are driving the whole problem.

But "the whole problem" goes deeper than this. Secular reason, he says, has trained us no longer to think of ourselves as spiritual beings, or of the creation as a divinely sustained organism. He says that secular reason has encouraged us to think of the earth as a mechanism which we control through our rational and technological powers. And the latest and most virulently problematic form of secular reason and secular humanism is the ideology of the market as the only way to govern human affairs. He goes on,

I think the solution to this is to recover an understanding of human beings as a divine creation, and as a microcosm of creation. If we don't see our creation and all creation as gifts of God we are in danger of abusing one another and creation. So there is a deep spiritual connection between our refusal to see ourselves as spiritual beings made in God's image, and our abuse of God's creation, which we also stop seeing as a creation. Instead we see it simply as something to be engineered for our best use, our maximal comfort and maximal choice.

The Old Testament prophet Jeremiah, Northcott says, was the first ecological prophet in literary and religious history.

His core message is that the people of Israel were no longer treating land as a divine gift and therefore as God's creation, so that's why they lost the land during the Babylonian exile. They were enslaving their own people and seeing the land as their own possession. They had abandoned the "everlasting covenant."

Jesus also warned us about the dangers of greed. To me, one of his most powerful parables is the parable of the man who wanted to build bigger barns, but died even before he could get started. It's a story about the wickedness of greed and lust for power, and the willingness to store up an excess of God's creation in such a way that it actually becomes a source of death rather than life.

Northcott says that the church is now in the same place it was in the fourth century, when the church and Rome were in alliance:

You had an empire which generated a culture of elite luxury and wealth and immorality, and the church was in cahoots with it. So people like St Anthony and the Desert Fathers left the cities and went into the desert to try to recover space and contemplate with God, and struggle to recover the soul they were in danger of losing in this collusive relationship with empire. The early monastic movement, under the leadership of St Benedict, was a further development of this quest to be true to the radical call of the gospel.

I believe the church is now in a similarly compromised place with the empire of our day; that we have this collusive relationship with the powers that be, and the powers that be in the world are increasingly capitalistic and technological and treating the earth as a mechanism and not as a creation.

However, there are groups of Christians, both here and in many parts of the world, who are trying to find a new way of living with some kind of common rule of life, growing their own food, gathering rainwater, living off renewable energy and having a common rule of worship.

I find that a very interesting development because it was out of the monastic movement that the church not only survived the collapse of Rome, but it also became the source of the birth of the European civilisation. And it seems to me we are again in a situation where we need another Benedict. Who will that Benedict be?

28. Is Earth 2.0 really "bad news for God"?

By Mark Lindsay

The discovery of earth-like planets does not disprove God's existence.

Since the discovery in late July 2015 of Kepler-452b—more prosaically but somewhat inaccurately known as "Earth 2.0"—scientific and non-scientific communities have been abuzz with excited speculation. The existence of another earth-like planet outside the solar system (an "exoplanet") in the so-called Goldilocks Zone has reignited interest in that perennial question: are we alone in the universe? It has also raised the intriguing possibility that, if there are other habitable planets "out there," we may one day be able to re-locate human civilisation to one of them.

These sorts of possibilities, for so long relegated literally to science fiction, have suddenly become credible. In 2010, the director of NASA's Ames Research Center, Simon Worden, predicted the availability "within a few years" of a spaceship "that will take us between worlds." Similarly, the Cambridge theoretical cosmologist Stephen Hawking has long insisted that inter-planetary re-location is the only viable option for humanity's continuing existence. And while technological barriers are at present insurmountable, NASA engineer Adam Steltzner believes that he will see human footprints on Mars in his lifetime.

But it is not only questions of scientific possibility that have now been raised with fresh enthusiasm. Earth 2.0 and the plethora of exoplanets discovered since then also pose questions that go to the heart of religion rather than science. For Jeff Schweitzer, former presidential adviser on science and technology to Bill Clinton, Earth 2.0 represents the "worst possible news for God" and all who believe in him. (Curiously, while he thinks that this will necessitate a re-write of all the world's major religions, he does seem to be particularly determined to focus attention on what he calls the god-myth of the Hebrew and Christian Scriptures.)

Whenever we consider space and all its possibilities, it's easy, of course, to be dazzled by the sheer numbers involved. Some estimates put the total number of stars in the observable universe at 10^{24}—that's a septillion (1 followed by 24 zeros) stars!

But for Schweitzer, the numbers are in fact irrelevant—it doesn't require billions, or even millions, of other possibly habitable planets. There only

Is Earth 2.0 really "bad news for God"?

needs to be *one* other, to disprove religious myth. In other words, even if Kepler-452b were the only other habitable planet besides earth, it would by itself be sufficient to discredit the idea of God (and specifically, the God of the Jewish-Christian Scriptures). Not surprisingly, the announcement of its discovery was therefore pregnant with both scientific and theological significance.

But before we rush to abandon our faith, it's worth considering the basis on which Schweitzer proclaims (again!) the "death of God." Frankly, there is no kind way to put it—he really should have stuck to science, about which I am sure he knows a great deal. Unfortunately, he seems to know next to nothing about textual exegesis. I'm an historical theologian by training, not a biblical studies expert—but even I can tell that his argument that Kepler-452b definitively disproves God rests on the thinnest of exegetical wafers.

Put simply, Schweitzer wants us to accept that because the Genesis creation narratives speak of life in the earth-bound singular, any proof or even possibility of *extra*-terrestrial life falsifies the Bible's own claims about the God who "instructed" its authors. If there were other lives or habitable planets apart from Earth, God would surely have told the biblical writers to mention them. Because he didn't, the Bible is therefore false and its God non-existent.

To prove his thesis even more definitively, Schweitzer then resorts to the age-old "straw man" line of argument. He uses Pope Urban VIII's condemnation of Galileo's heliocentrism in 1663 to demonstrate beyond doubt that the church, its Bible, and in consequence the God it worships, is demonstrably ridiculous and not worthy of being taken seriously, particularly on scientific issues. (Schweitzer conveniently ignores the fact that in 1992 the Vatican formally acknowledged that Galileo had been correct.)

Of course, at the most basic level, Schweitzer is right. The Bible does only concern itself with the creation of life on this planet. For all its talk about "the heavens," there is no mention of other possible earth-like worlds, nor do any of the writers enquire into the existence of alien life-forms. But is this really the fatal oversight that Schweitzer believes it to be? Why should the texts, even those that speak of God's astronomical creativity, necessarily go beyond the horizon of this-worldly experience? That is hardly their purpose. As the British Rabbi Abraham Cohen once noted, "natural science as a subject of study was not cultivated in the schools of Palestine or Babylon." The rabbis were aware of, but largely uninterested in, cosmology as such. Why? Because they were sufficiently concerned with the problems of *this* world, without worrying about speculative possibilities of any others. As a famous statement from the Mishnah puts it, "Whoever reflects on four things, it were better for him that he had not come into the world: 'What is above? What is beneath? What is before? and What is after?'"

That is to say that the Bible is simply uninterested in the maybes and what-ifs of astrophysical abstraction. On the contrary, the Bible in its entirety provides a narrative framework for understanding how *this* people (the community of Israel and then, later, the church of Jews and Gentiles) relates to *this* God. There is something deeply grounded about Scripture's intention—to not enquire abstractly into what may be, but to ask instead, how then should we live, here and now, with ourselves and with one another? To require of the Bible that it also engage in cosmological speculation is as nonsensical as demanding that we find a recipe for butterscotch duff in Einstein's general theory of relativity, or the clue to a crossword puzzle in Newton's *Arithmetica Universalis*.

This is not, of course, to deny that the discovery of Earth 2.0 and others like it is freighted with theological significance. No longer living in the time of Scripture's composition, we do not have the luxury of ignoring the scientific realities that were, at one time, mere curiosities. Theologians, religious leaders and people of faith generally, should feel inwardly compelled by such realities to ask probing questions about the nature of the God in whom we say we believe.

Such questions might include: What does it mean to say that God is pro nobis (for us)? Is he also "for" all other possible life-forms? (Why not, we might ask?) Is there still something unique about life made "in God's image" that may not of necessity be true of any other life-types? If we already confess that God is the creator of "the heavens"—which is itself a Scriptural term—can we in good faith also affirm that God is the maker of "the earths"? Perhaps particularly, is there anything in the Bible that would prevent us from making such a confession?

Let me hasten to add that none of these questions requires us to abandon our belief in the Scriptures of the Old and New Testaments as the revealed word of God, in which we find "all things necessary to salvation." They are, it seems to me, no more problematic for the nature of biblical revelation than the questions posed of the Bible by the historical criticism of the nineteenth century.

But they are questions that we should ask carefully, and in respectful dialogue with the appropriate scientific disciplines. To do so would be

Is Earth 2.0 really "bad news for God"?

nothing other than what was proposed by the Jesuit theologian Bernard Lonergan, in his ground-breaking book *Insight*. But in all honesty, asking such questions and engaging in such dialogue is in fact merely the natural extension of what the various process and eco-theologies have been doing for decades. Eco-theology has intentionally sought to destabilise the anthropocentric, geocentric narrative of normative doctrine; to ask now about the impact of *extra*-terrestrial habitation on our concepts of God is different only by degree and not by kind.

But if we are to ask these sorts of questions, and I believe that we must, we should be guided by a far more sophisticated exegetical grammar than that employed by Jeff Schweitzer. We will need to learn how to read our foundational religious texts with nuance, and with an eye to the discursive whole, rather than contenting ourselves with facile proof-texting.

One thing is for sure. Theology—just as much as science—needs to keep an open mind and an open heart. If it does so, then that will be in fact good news for God, or at least, for our proclamation of him.

This chapter is an expanded version of an article which first appeared on *The Conversation*. See http://theconversation.com/au

Minds, Brains and Spirituality

29. Meditation, mindfulness and the brain

By Barry Rogers

There is growing scientific evidence of the positive impacts of regular, repeated meditation and prayer on the brain. Barry Rogers includes meditation, mindfulness practice and gratitude exercises in his psychology practice.

When working with anxious, depressed and stressed clients, while diagnostic assessment and symptom relief is an important part of their therapy, personal wellbeing increases when people can improve awareness of their thoughts and identify personal strengths for increased resilience. Therapy includes basic meditative exercises, including slow breathing, cognitive reframing, mindfulness practice, a review of character strengths, and a gratitude exercise. These reflective practices help reduce anxious and depressive symptoms while also encouraging personal awareness and compassion for others.

Introducing some basic brain–body biology helps increase awareness of "where they are in the brain." Are they regularly in a reactive/catastrophising space (the amygdala) using an anxious or pessimistic thinking style; or in the prefrontal cortex space, the organising/reviewing/initiating/managing region of the brain? The amygdala, in the temporal lobe regions, is where emotions and emotional behaviour and motivation are reviewed and integrated around aspects of survival, in a brain area that includes emotional learning and memory processing. In times of perceived stress, the amygdala response is their first port of call, with repeated anxious or pessimistic perceptions occurring about life issues and people. A form of "negative chaining" can occur, linking negative or worrying thoughts together without mindful review, establishing reactive thinking patterns/schema in the brain. So the ways in which we think and reconstruct our reality, based on memories and emotions linked to our experiences and relationships, can either promote or hinder wellbeing.

Meditation and prayer are important here, directing us towards mindful emotional learning: reviewing the feelings and intentions of others and ourselves, which helps to develop well-being capacity. For those who have hard-to-shift memories around perceived needs and difficulties arising from developmental experiences, these negative outcomes can be processed, reprocessed, and reinforced. Since the amygdala and brain stem (the latter is also activated during trauma) are automatic response systems linked to the body's autonomic nervous system, a lack of engagement with these default thoughts tends to ensure that bodily reactions and symptoms associated with anxiety, depression, and stress (including trauma) will continue regularly, unabated, until the body eventually comes to rest through this emotional and cognitive "workout." What we attend to currently, then, can either promote or block wellbeing. The perceptions we hold are what we have "in mind": what we're mindful of.

In Scripture, Paul provides some insights—and antidotes—to anxious reactivity, and is a good example of someone who learned to counter stressful triggers after his more traumatic and difficult experiences as a high-profile Christian disciple (Acts 9:23-30; 14:19-20). His "dark night" experience of Christ (Acts 9:1-19) included a temporary loss of vision, so that he has to use other faculties to make sense of his Christ-encounter, re-framed from his previous life-outlooks. These had included angry, vengeful, judgemental, pessimistic, reactive outlooks and behaviours towards the early Christian "sect" as he perceived it (Acts 8:1; 9:1-2). Years later, and following his Nabatean/"desert" experience (Gal 1:17ff), we find a less combative Paul, who, despite his suffering, displays a sense of contentment through Christ, and is others-focussed, in a compassionate way. In Philippians 4:4-7, he counsels towards a positive displacement *away from* anxiety-based thinking to gratitude and thanksgiving to God, where the wellbeing (*eirene/shalom*) from God will "guard your hearts and minds *in* Christ Jesus."

This displacement is towards dynamic virtues that reflect wellbeing: "Whatever is true/honourable/just/pure/pleasing/commendable/of excellence/worthy of praise—*think about* these things" (Gk. *logizesthe*—think upon, value, reckon—used originally in a financial sense—what is *valued*, 4:8). What we value or give attention to, in a mindful or present focus towards prayer and meditation, is where the brain will "go." Its neuronal structures and pathways will regularly "fire" in this capacity, rather than in anxious or pessimistic responding. When we pray and meditate, there is a positive neuronal shift in the brain that reduces stress. We might ponder Psalm 119:15: "I will meditate in thy precepts, and have respect unto thy ways" (Mamre Institute translation); a scriptural phrase such as the "Maranatha" prayer popularised by the late John Main; or the Jesus Prayer ("Lord Jesus, have mercy on me"), in seeking Christ's eternal presence (1 Cor 16:22; Rev 22:20).

Meditating has been shown to trigger brain activity that over time creates regular, positive changes in the growth and distribution of grey matter in the brain, promoting wellbeing. This occurs in the lateral prefrontal cortex, the brain region that allows us to review things with a more rational, logical, and emotionally balanced perspective: what psychiatrist Rebecca Gladding calls its *assessment centre*. Other recent research has also identified the nearby medial prefrontal cortex which, with support from the hippocampus, "likely forms and stores schema [neuronal structures] which map context and events onto appropriate actions. ... These schema purposefully direct the correct emotional or motor response to a given set of events in light of past experience" ("The Role of the Medial Prefrontal Cortex in Memory and Decision Making," Euston at al., *Neuron*, 2012).

Regular, repeated practice of meditation and prayer engenders memory constructs that are very different to those that regularly, or repeatedly over time, focus on resentment, trauma, or difficult relationships where there has been little or no resolution of past concerns. Meditative prayer and mindfulness practice, while having somewhat different foci, have similar positive effects in brain activity. Focussing on *practice* is the key here. A mindful focus on what is true, honourable, etc., strengthens the neuronal response pathways, which impact positively on the body, in lowering heart and breathing rates, reducing external and internal muscle and body tension, and improving the ability to "collect our thoughts"/focus, leading to better insight. These are polar opposites to the outcomes associated with negative stress.

A growing body of evidence

Neuroscientist Dr Judson Brewer started practising meditation during his medical training. With his research colleagues he recently identified long-term benefits of meditation, based on MRI brain scans of experienced meditators (those meditating for 10 years or more). In the range of meditation practices, they observed "decreased activity in the areas linked to attention lapses, anxiety, attention-deficit hyperactivity disorder, schizophrenia, autism, and plaque build-up in Alzheimer disease" (*Yale Medicine*, 2011). The brain regions scanned were the medial prefrontal and posterior cingulate cortices referred to earlier.

Another review of 47 independent trials (3,515 participants), identified mild to moderate health benefits from mindfulness and mantra meditation techniques in reducing anxiety, depression, and also pain ("Meditation

Programs for Psychological Stress and Well-being: A Systematic Review and Meta-analysis," Goyal et al., JAMA *Internal Medicine*, 2014). It also concluded no harmful effects from meditation practice as a therapeutic alternative, with effect sizes similar to those for prescribed medication, in reducing and managing incidences of depression.

Another research report on the psychobiology of depression identified the positive effects of regular mild physical exercise, daylight exposure, *mindfulness* and a good hydration and eating regimen in reducing depressive symptoms ("How to Increase Serotonin in the Human Brain Without Drugs," (Young, *Journal of Psychiatry & Neuroscience*, 2007). Mindfulness meditation is now one focused psychological strategy for addressing moderate to severe depression, increasing positive mood and overall wellbeing via serotonin shifts in the brain.

Neuroscientist Sara Lazar and her team at Harvard University found that mindfulness-based stress reduction changes brain structure. Her 2011 TEDxCambridge talk reported the team's findings on structural changes to the brain for participants in an eight-week meditation course on mindfulness-based stress reduction. Neuroimaging techniques were used to review neurological, cognitive and emotional changes, and physiology via breathing and heart rate, to understand brain-body interactions during meditation practice. Meditators showed increased cortical thickness in the hippocampus, the brain area governing learning and memory, and in areas of the prefrontal cortex noted earlier. They also showed a *reduction* in the amygdala's brain cell volume, correlating with observations of reduced anxiety, fear and stress, and *increases* in grey matter in their prefrontal brain areas compared to control subjects (those who didn't meditate).

This outcome reflects the central idea behind neuroplasticity: that the brain can be "trained" away from non-beneficial behaviours and thinking to those that are: "the neurons change how they talk to each other, with experience" (Lazar, 2011; see also Doidge, *The Brain That Changes Itself*, 2007). Increases here in cell matter carry positive effects via decreased stress and reductions in common mental health symptoms for depression, anxiety disorders, pain and insomnia, as well as an enhanced ability to focus and some overall increases in quality of life. Positively, brain neuroplasticity is neither area nor age-specific. Lazar's research showed that while cortical areas routinely shrink with ageing, these effects are lessened through meditation: "The 50-year-old meditators had the same amount of cortex as the 25-year-olds, suggesting that meditation practice may actually slow down or prevent the natural age-related decline in cortical structure."

In a subsequent study, non-meditators had initial MRI brain scans, and then regularly meditated for 30–40 minutes within an eight-week stress-reduction program. Their MRI scans after the program showed *increases* in grey matter in the hippocampal region (memory and emotion regulation) compared to controls. And the temporo-parietal junction (linked to perspective-taking/discernment, empathy and compassion) also showed increases in grey matter. By contrast, their amygdalas had *reductions* in grey matter, correlating with the reduced stress being reported. Nothing else had changed in other areas of their lives: work responsibilities, other stressors and previous challenges remained. There were also positive changes in outlook, including an increased sense of compassion. In short, meditation improved brain functioning for personal and social resilience in life.

Gratitude and thanksgiving are central

These are significant findings for Christian experience when we recall Paul's remarks contrasting the spiritual debility outlined in Romans 1–7 with his concluding affirmation of *no condemnation for those in Christ* (Rom 8:1). In meditative prayer we experience a *renewal of the mind* in engaging with the Spirit, who sets the mind in a prayerful focus: "help[ing] us in our weakness ... interced[ing] with sighs too deep for words" (Rom 8:26–27).

A central, prayerful focus in Paul's letters is gratitude and thanksgiving because of what God has accomplished for us in Christ. We see this also in the thanksgiving Psalms (e.g., Ps 138) and Jesus' thanksgiving prayer for his followers (Matt 11.25).

Gratitude has also been identified in the research-based classification of "Values in Action (VIA)," a constellation of 24 human strengths including *spirituality, sense of purpose, and faith* (Peterson and Seligman, *Character Strengths and Virtues*, 2004). They identify motivations for hopeful, optimistic engagement with others in our relationships and work contexts.

Since 2008, the VIA strengths review has been included in the assessment and mentoring process for those undertaking ordained ministry training in the Melbourne diocese. An extended *gratitude* exercise is also part of the Masters in Positive Psychology program at the University of Pennsylvania. The "three blessings" exercise students undertake recalls the well-known hymn (1897) of Johnson Oatman Jr, which focused on the character strength of gratitude as a pathway to resilience: "When upon life's billows you are tempest-tossed, When you are discouraged, thinking all is lost, Count your many blessings, name them one by one, And it will surprise you what the Lord has done."

A key understanding in Christian experience is that Christ prays on our behalf, as he did before the cross (John 17), so that we may have life in the Spirit. This is where the renewing of our mind begins, and continues, as we engage regularly in meditative prayer that includes adoration and gratitude, in focussing on Christ's presence with us.

30. Psychical research a wonderful gift to religion

By Colin Goodwin

Paranormal phenomena such as extra-sensory perception and near-death experiences make a compelling case for believing that being human and our universe cannot be reduced to just the material.

The renowned American psychologist J. B. Rhine (1895-1980) never hesitated to refer to psychical research or parapsychology as "religion's science." What he meant was that psychical research offered highly disciplined empirical investigation into certain phenomena that went beyond our normal expectations and causal accounts of things, and that, by reason of its content, drew attention to the limitations of the sort of materialistic interpretation of the human person and of the universe so widespread in our modern world—and so totally inimical to religion. We have here an unparalleled linking of science (understood as a systematic, experientially grounded, exploration of some or other part of reality) and religion.

What are these paranormal phenomena meticulously studied in psychical research that tend to undermine philosophical materialism, and open up the possibility of a religious or spiritual worldview? These phenomena fall into three main categories. One category is *extrasensory perception* (ESP). Within this category is clairvoyance (the direct perception of physical objects without the aid of normal perception by our sense powers), telepathy (the direct communication of thoughts, feelings, intentions, and so on, from one mind to another independently of any recognised channels of sensory communication, e.g., speech, writing, or gestures), and precognition (clairvoyance relating to an event or state of affairs yet to come about, e.g., a future train accident).

Clearly the occurrence of clairvoyance, telepathy, and precognition, seriously challenges the correctness of materialism's depiction of the human mind, consciousness, and personality, as being one-sidedly and completely dependent on the neuronal workings of the central nervous system of brain and spinal cord.

A second category of paranormal phenomena is *psychokinesis* (PK). Psychokinesis is direct control by the mind of material objects other than certain parts of one's own body without the use of any physical means whatsoever. Typically PK involves the deliberate moving of objects such as selected matchsticks across a table, mind-initiated moving of objects into or out of sealed containers, the influencing by thought only of the fall of dice to give pre-intended numbers, the bending by thought alone under controlled conditions of keys or spoons (a feat sometimes accomplished by children as young as ten). The notion of PK may be extended to include the activity of poltergeists (German for "noisy spirits") which, by exercising mind only and using no physical means at all, startlingly disturb what is around them, e.g., by moving cabinets, making items fall from shelves, pushing people down flights of stairs, and so on.

Analogous to ESP, PK reveals powers of mind that both surpass material structures and influence a physical world open to these powers—a further throwing of doubt on the adequacy of materialism as a worldview.

The third category of paranormal phenomena comprises experiences that point to the capacity of minds or psyches to exist apart from their physical bodies. Well attested out-of-body experiences, near-death experiences, and messages through mediums, offer varying kinds of evidence for this capacity. The capacity of human beings for soul-survival of bodily death is evidenced also from pre and post mortem apparitions of particular individuals, the haunting of specific locations by disembodied souls or selves of deceased persons, and (most surprisingly) the searchingly investigated, extensively documented, cases of reincarnation, that is, of the entry into new human bodily existence of the souls or selves of some people who have died. (The work of the late Professor Ian Stevenson, Head of the Division of Personality Studies, University of Virginia, is outstanding in this connection).

The reality of this third category of paranormal phenomena decisively contradicts a materialistic account of the human person, and of the universe of which the human person is a part.

The published testimony supporting the reality of paranormal phenomena within the three categories identified above is simply immense. There is space here only to mention the sustained collating (and critical) work of the Society for Psychical Research, founded in England in 1882 by a trio of Cambridge academics; the comparable work of the American Society for Psychical Research, founded in 1885 under the aegis of the extraordinary William James (who once bluntly commented in a scholarly journal, "The concrete evidence for most of the psychic phenomena under discussion is good enough to hang a man 20 times over"), and of the J. B. Rhine inspired American Parapsychological Association; and to note that psychical research has been carried out, and carefully reported, across

Psychical research a wonderful gift to religion

Europe for well over a hundred years, and in more recent decades in Russia, India, and China.

Perhaps predictably the scientific establishment is loath to include psychical research within its remit. Nobel Prize winning Cambridge scientist Brian Josephson is critical of this reluctance: "Physicists have an emotional response when they hear anything connected with parapsychology. Their opinion ... is not based on evaluation of the evidence but on a dogmatic belief that all research in this field is false." Josephson adds pointedly that scientists' "irrational attacks on parapsychology" arise out of the sharp challenge of "putting these phenomena into our present system of understanding the universe."

Endorsing, and going beyond, the Nobel Prize winner's words is the stated aim of Duke University's Rhine Research Centre for parapsychology: "to improve the human condition by creating a scientific understanding of those abilities and sensitivities that appear to transcend the ordinary limits of space and time." Against a background of acknowledging such "abilities and sensitivities" one may understand the emergence in England in 1953 of an ecumenical group of clergy and laity to "promote the study and integration of psychical and spiritual experience within a Christian context"—a group that evolved into the Churches' Fellowship for Psychical and Spiritual Studies (CFPSS), operating under the important motto "To faith add knowledge." Papers presented at CFPSS annual conferences, or read at its regional meetings, provide compelling proof of wide-ranging, up-to-date, scientific study of paranormal phenomena that effectively closes gaps between science, on the one hand, and religion, on the other. CFPSS is a model for "viewing the paranormal through a scrupulously honest Christian lens," as noted English author Susan Howatch, a patron of CFPSS, once remarked.

Psychical research is a wonderful gift-horse from the field of the sciences. Religion would be well advised not to look it in the mouth.

31. Religious sense at the heart of self-perception

By Andrew Wood

Is there evidence for a "God spot" in the brain?

Medical imaging systems have shown remarkable advances in recent years. The ability to image the workings of the brain while a person makes moral and ethical decisions or whilst they contemplate works of art has extended the range of investigations from purely diagnosing illness to fundamental studies of the ways different people think. It has meant that topics which were previously thought to be "off-limits" to science (beliefs, emotions, moral reasoning) are now increasingly the objects of systematic experimentation. Bluntly, neuroscientists are seeking to find if there are specialised brain regions where people "do" religion, both in terms of beliefs and behaviours.

What makes all this possible? Briefly, the trillions of nerve cells within the brain communicate by passing tiny electrical currents around themselves. If a particular part of the brain is engaging in a task (i.e., "thinking") then the currents may be a little higher in that region. Thus, by measuring the effects these currents produce on the scalp, a "map" of brain activity can be recorded by using a grid of electrodes. This forms the basis of electroencephalographic (EEG) mapping. Since nerves expend energy after producing these currents, there is a local consumption of such "fuels" as dissolved oxygen and glucose (or sugar). Other techniques build up a three-dimensional map of the brain, sensing those regions where more than average amounts of oxygen or glucose are being consumed. The names of these techniques are functional Magnetic Resonance Imaging (fMRI) for oxygen and Positron Emission Tomography (PET) for glucose. The advantage of EEG over these newer techniques, is that very small differences in time between brain events can be recorded, but, on the other hand, it is poor at localising events, the other techniques being superior.

Precision is improved by comparing two images, taken when the person is responding to two different types of task. The differences between the two brain images can then be connected to the differences in the tasks. The tasks can be responding to statements, flashed on a computer screen, such as "The biblical God really exists" or "Santa Claus really exists." The reaction of a believer to these two statements is expected to be different, whereas for

Religious sense at the heart of self-perception

a non-believer might be the same. From the brain maps, the regions of the brain associated with religious, and maybe specifically Christian beliefs can be identified, by noting the local strength of differences between believers and non-believers. Investigators use previous findings to discover what other traits or behaviours are associated with the same regions. They then use this knowledge to link religious behaviour with other behaviour. For example, more than one study has identified a brain region associated with "sense of self" with acceptance or rejection of religious statements. However, by the nature of the questions, people who are religious may take longer to evaluate religious compared to non-religious statements, because these beliefs are important to them. Thus it may not be a fundamental difference in brain architecture (a "God spot") giving rise to these differences. The evidence points to the same brain processes being involved in deciding on whether to believe or disbelieve in religious propositions as are used for general propositions.

Religious experience rather than religious belief has also been investigated using these methods. For example, brain images obtained while religious people have been reciting prayers, "speaking in tongues" or during religious contemplation have been compared against more mundane activities. In general, these are less than satisfactory in relation to making a valid comparison. The MRI machine, for example, is noisy and claustrophobic and not the environment naturally conducive to a sense of closeness to God. Even EEG mapping involves the attachment of maybe over 100 electrodes to the scalp. Nevertheless, comparisons were made with a group of Carmelite nuns who were asked to recall, within the MRI machine, the most intense religious experience they had ever had. Their brain activity whilst they were doing this was compared to that when recalling an intense personal relationship. Not unexpectedly, the regions which were differently affected, were those to do with "sense of self." Our religious sense is thus at the very centre of how we perceive who we are.

Similar experiments have been carried out during the activity of praying, with studies carried out in a number of places, including Denmark, where again fMRI techniques were used with participants engaged in formal praying and improvised praying compared against reciting nursery rhymes and making wishes to Santa Claus, respectively. Whilst it could be argued that the comparison tasks could have been better selected, the improvised prayer was found to be comparable to a more ordinary personal conversation. Again, this may not seem surprising. More recently, the same group has studied Christian versus secular participants who were in receipt of (pre-recorded) intercessory prayer. The participants were told that six of the prayers were read by non-Christians, six by Christians and six by a Christian known for his healing abilities. In fact, all 18 prayers had been pre-recorded by "ordinary" Christians, so there was some deception involved. The slightly surprising finding was that when the Christian participants believed that the prayer was spoken by someone with well-known healing ability, they behaved in a way similar to people being hypnotised, "handing over" decision-making processes. In other words, the implicit trust in charismatic leaders (in this case imagined) led to a certain "switching off" of critical faculties. This is perhaps something that has long been suspected, but now techniques of neuroscience mean that it can be quantified.

"Speaking in tongues" or glossolalia in Christian experience has long been controversial, some regarding it as a uniquely Christian phenomenon and the mark of the indwelling of the Holy Spirit. It is not the place in this chapter to enter into a debate on the necessity of this phenomenon to be manifest in a live church nor whether it is an exclusively Christian one. Like other religious behaviours, this too has been subject to scientific investigation. Some have examined ecstatic utterances for similarities with and differences from conventional languages and have noted that it is not a language with syntax or grammar, nor is it for conversation, but that it nevertheless leads to personal benefits in terms of measurable improvement in mood and social integration. They have also noted that people become better at it with practice and thus it is perhaps a result of social learning rather than a semi-miraculous event. A study, with just five individuals, studied brain activation patterns during glossolalia compared with during singing. As in the study of the attitude to charismatic healers' prayers, glossolalia was associated with a reduction in the amount of intentional control, which is consistent with the notion of "letting go" during speaking in tongues.

The study of religious phenomena by neuroimaging methods has not been limited to Christian groups. Two recent studies have been of Buddhist monks, the first in China, the second in Italy. The first is interesting in that the two groups, Tibetan monks and Han Chinese, represent two groups with very different worldviews. In particular, the monks have a view that the "self" as experienced in the physical world is illusory, thus a minimal subjective sense of "I-ness." This was reflected in different brain activation patterns in relation to tasks involving contemplating the self compared with contemplating other people. The second was of two different meditative

Religious sense at the heart of self-perception

tasks: focused attention versus open monitoring. This showed that not all meditative activities are associated with "switching off" the will and that certain meditative activities show a great discipline of the mind.

In summary, mapping brain activity during religious and specifically Christian contemplation is now possible through advanced imaging techniques. Even the difference between lying and telling the truth has been measured in these types of experiments. Although there are deficiencies in the design of some experiments, the findings seem to confirm pre-conceived notions. The approach also assumes that brain activity is all there is to "thinking" and that regional brain activity can be tied rather rigidly to particular thought processes, neither of which is true.

Some might find the whole notion of Christian faith and practice being reduced to certain localised nervous activation patterns confronting, especially where there is a suggestion that similar patterns could be produced by implanted stimulating electrodes, or could be obliterated by a local injury. However, the evidence does not support this extreme reductionist view. It should be emphasised that there is a good deal of inconsistency between findings and there is a realisation among those who are honest that there is a great deal more complexity to brain processing than is identified in these types of experiments. Meanwhile even some of the most sceptical have to admit that rather than religion withering in the light of modernity, "it remains one of the most prominent features of human life in the twenty-first century." So, maybe we must face up to religious belief as being an innate feature of what it is to be human. In particular, when Jesus said "if you really knew me you would know my Father as well" he is referring to the realisation and embodiment of what this innate feature is pointing us towards.

The Human Story

32. Graeme Clark: A lifelong mission to bring hearing to the deaf

By John Pilbrow

A review of *Graeme Clark: The Man Who Invented the Bionic Ear*, by Mark Worthing (Allen and Unwin, Sydney 2015)

A well-written and very readable biography of a great Australian, Professor Graeme Clark AC, this is a fine portrait of a humble yet very determined, highly qualified and widely experienced medical scientist who played such a pivotal role in the development of the bionic ear. It complements Clark's autobiography (*Sounds From Silence: Graeme Clark and the Bionic Ear Story*, 2000) by revealing more about the man behind the story.

It was his father's deafness that led Clark to resolve at an early age that his life's work would be to make a difference for the deaf. Graduating in medicine from Sydney University, then qualifying in surgery in both Edinburgh and London, Clark returned and practised mainly nasal surgery in Melbourne. However, to follow his passion to restore hearing loss, he enrolled in a PhD at Sydney University so as to explore what was known about electrical stimulation of the inner ear.

Melbourne University appointed Clark at 34 as the first professor of otolaryngology in the country, based at the Royal Eye and Ear Hospital in East Melbourne. Here he gradually assembled a large team of talented researchers in medicine, electronics, bioengineering, computing, speech processing and recognition. Not all was plain sailing. A good deal of opposition to his proposed bionic ear came from professorial colleagues. Some thought he was mad to try, others that it was impossible. Also there were many in the deaf community who felt he was undermining their communities. But he persisted.

By the mid-1970s one critical problem remained. Could a bundle of electrodes inserted into the inner ear enable the recipient to recognise speech? I well recall the occasion around 1974 when several of us, including Graeme Clark, met visiting British neuroscientist Donald Mackay to discuss science and faith. Impressed by what Mackay had to say, Clark later arranged a six-month sabbatical with him and his group at Keele University where he gained critical new insights into speech processing.

The Human Story

Clark's adult Christian journey began when he made a commitment to Christ during a Student Christian Movement (SCM) camp in the 1950's while at Sydney University. He acknowledges the many talks to the SCM by the late Charles Birch that helped him to understand biology and evolution. During his return to Australia from surgical training, several encounters with Christians, two of them fellow medicos, nudged him towards an evangelical position. Conversations with Donald Mackay during his sabbatical at Keele also played a role. Clark's Christian journey has been underpinned by prayer; on many occasions he has gone away to pray, particularly at critical points in the bionic ear journey.

Fully supported by his wife, Margaret, Clark was always committed to balancing his professional and family life and he richly deserved being chosen as Australian Senior Father of the Year in 2004.

The moment of truth arrived in 1978 when Clark operated on Rod Tucker, a man whose hearing had been lost in an accident. This was the first successful cochlear implantation in the world. An equipment fault caused the first post-operative test to fail, but when repaired, "God Save the Queen" was played and Tucker jumped to attention, pulling out many of the electrical leads!

Clark has received many honours including Fellowship of The Royal Society of London, Fellowship of the Australian Academy of Science and he is only the third Australian to receive the Lister Medal of the Royal College of Surgeons. He has received several honorary degrees, was Senior Australian of the Year (2001) and ABC Boyer Lecturer (2007). Along with three other cochlear implant pioneers, Clark shared the Russ Prize from the US National Academy of Engineering (valued at $US500,000).

The biography was launched in 2015, at Melbourne's State Library, by Sian Neam-Smith, at one time the youngest-ever recipient of a bionic ear at only two years old. She received a second bionic ear at age 19. Sian is a very articulate university graduate, and to hear her speak one would not have known that she had been born deaf.

The foreword by Li Cunxin, author of *Mao's Last Dancer*, includes these words, "I found Mark Worthing's story of Graeme Clark moving and inspiring. It captures the spirit of a truly amazing man who changed the life of our daughter and of many thousands of others in the world."

33. Jürgen Moltmann: How "Christ the bridge" led me from science to theology

By Jürgen Moltmann

Jürgen Moltmann describes his journey from being a non-religious young soldier and aspiring scientist in Nazi Germany, to a POW in the UK, convert and theologian. Stirred and shaped by his experiences of death and suffering, he explains how he came to discover "his little life story" in the story of the passion of Christ, and explores how the search for truth, knowledge, wisdom and beauty involves both science and spirituality.

I was never a learned scientist. I was not even a student of physics. But science is for me a "paradise lost." Chemistry, physics and mathematics formed the dreamland of my youth. I was part of a young scientific community in school. We tried chemical experiments in hidden cellars; in friendly rivalry we told one another our newest findings in atomic physics. We were a group of schoolboys training our minds in science not only to impress our teachers. It was a wonderful time of awakening of the spirit of discovery of the truth and beauty of nature.

The end came early. In 1943, just as I was reading a new book by the French scientist Louis de Broglie, *Matiere et Lumiere*, published in German with a long foreword by Werner Heisenberg, I was called up and conscripted into the German army. I was sixteen years old. In Hamburg the school classes were distributed to the anti-aircraft batteries of the city. We were made into "Luftwaffenhelfer" (Air Force auxiliaries). I was sent to a battery in Schwanenwiek, that stood on stilts on Hamburg's large lake, the Outer Alster. It was probably on stilts to allow a wide firing range in all directions, but it also made the battery very visible from above. We already felt like "soldiers"; teachers however came during the day and taught us over-tired "warriors" who had been on the alert during the night. It was here that Dr Magin, a teacher of mathematics, found me. I shall never forget him; he was only interested in interested students and didn't care about the rest of the class. My friend Gerhard Schopper and I became his devoted disciples.

In July 1943 the Royal Air Force began the bombing of Hamburg. For a whole week several hundred bomber squadrons were over the city. The eastern part of the city was completely destroyed as more than 30,000 people died. During the final night of the raid our battery was bombed and destroyed. One bomb hit the platform where we were standing. A mass of splinters destroyed the firing device and Gerhard Schopper, who was standing next to me, was torn apart. He hadn't got down quickly enough. I stood up blinded and deaf, but only slightly wounded. That night, for the first time in my life, I cried out to God: "My God, where are you? Where is God?" Then I was tormented by the question: "Why was I not dead too, like the friend at my side? Why must I live?" I came from a secular family and was not religious, but in that night of terror and death my "wrestling with God" began.

"Operation Gomorrah": that was the name the biblically versed Air Force commanders gave their first planned destruction of a big German city. When we heard this we were horrified, until we realised that during the Nazi regime more than 20,000 people were put to death in the Neuengamme concentration camp near Hamburg, and that Hamburg Jews were murdered in White Russia; Hamburg was no innocent city. As a survivor of "Gomorrah" I am a survivor of these Hamburg catastrophes too and feel "guilty" and duty bound to the dead because of my survival.

In my younger years I had wanted to become a scientist; after five years of war and captivity I came back a theologian. Questions about God had become more important to me, although the questions about scientific truth have always remained with me. I shall first tell you about my journey to theology, because I am indebted to Britain for this. In the second part I shall deal with the slogan "knowledge is power" and the idea that science is nothing more than a way of seizing power over some parts of nature, and confront this idea with the other ideas: truth is beauty and knowledge is wisdom and wisdom is the ethics of science. In conclusion I shall deal with the question: How far does the responsibility of the scientific community go in questions of war and peace of our nations and in the life-and-death questions of humankind?

The long road to theology

My personal journey to theology was undertaken in prisoner-of-war camps in the UK. In 1944 I was drafted into the German infantry. We were briefly and poorly trained before being sent to the Netherlands in September 1944. British airborne troops had landed in Eindhoven and Arnhem and we were sent to stop them, though we were more the hunted ones than the hunters. Then in February 1945 a British-Canadian operation through the

Jürgen Moltmann: How "Christ the bridge" led me from science to theology

Reichswald began. My unit was scattered, I lost my way at night and found myself alone in the woods, until I met a group of British soldiers. I called "I surrender," and they didn't shoot me. The next morning a compassionate lieutenant gave a plate of baked beans to the hungry prisoner. I have loved baked beans ever since—for me, they taste of life!

My first POW camp at Zedelgem near Ostende in Belgium was quite miserable: 200 men in an old ammunition shed with the threat of Nazi terror attacks at night until the end of war. Afterwards we just felt numb with despair as we learned that most German towns had been destroyed, that twelve million displaced people had been expelled from Silesia and East Prussia ... but no personal news came through. We had escaped death, but we had lost all hope. The thought of there being no way out was like an iron band constricting the heart. And each one of us tried to conceal his stricken heart behind an armour of untouchability.

The worst however took place in one's soul: every night the nightmares of war, terror and death returned. The faces of the dead appeared and looked at me with sightless eyes. Every night it was like wrestling with the dark side of God. In those nights one was "alone" like Jacob at the Jabbok river fighting powers that seemed dark and dangerous. It was only afterwards, and later, that it became clear to me with whom I had been wrestling. It was three years at least before I found some healing for these dangerous memories.

We had escaped the mass death of the world war. But for every one who survived hundreds had died. One had to bear the weight of grief. My spiritual nourishment had been the poems of Goethe and Schiller. I knew many of them by heart. They had awakened and formed my boyhood emotions, but now, shut in with 200 others, I found they had nothing more to say to me. My dream of science faded.

What was the point of it all? I lost interest in life. All my senses were dulled: I had eyes, but saw nothing, ears and heard nothing. I felt nothing any more. Then something unforgettable happened to me. In May 1945 as we were pushing a truck out of the camp I found myself suddenly standing in front of a cherry tree in full bloom. Life in all its fullness was before me. Deprived as I had been of any interest in life, I almost fainted with the overwhelming joy of it. I saw colours again and sensed life in myself once more. The Spirit of Life had touched me.

In August 1945 my name was called and we were taken to a ship in Ostende, sailing not for Hamburg but to London. The next morning we passed under Tower Bridge. By train we reached Scotland and ended up in Kilmarnock, Ayrshire. We had been transferred to an old, well-equipped POW camp, where only twenty of us shared a Nissen hut. After a variety of jobs I advanced to become interpreter for a working group on the streets of New Cumnock. The Scottish overseers and their families were the first to come into friendly contact with us. They met us, their former enemies, with a hospitality that profoundly shamed us. We heard no reproaches, we were not blamed, we experienced a simple solidarity and a warm common humanity. For me this was quite overwhelming. They made it possible for us to live with the guilty past of our people without repressing it and without growing callous. True, we had numbers on our backs and prisoners' patches on our trousers, but we felt accepted. That humanity in far off Scotland made human beings of us once more. We were able to laugh again.

In September 1945 we were confronted in the camp with pictures of Bergen-Belsen and Buchenwald—concentration camps. They were pinned up in a hut, with only laconic commentaries attached. Some of us thought it was just English propaganda. Others set the piles of dead bodies against the destruction of German cities. But slowly and inexorably the truth seeped into our consciousness, and we saw ourselves through the eyes of those Nazi victims. Was my generation the last to be driven to death so that the concentration camp murderers could go on killing people? For me every patriotic feeling for "Germany the Holy Fatherland" collapsed and died. Depression at the extent of wartime destruction and captivity with no end in sight was compounded by a feeling of profound shame at having to shoulder a share of the disgrace of one's own people.

One day a well-meaning British army chaplain came to our camp and after a brief address distributed some Bibles. I read the book in the evenings without much interest, until I came upon the psalms of lament in the Old Testament. In particular Psalm 39 caught my attention:
  ... I remained utterly silent,
  not even saying anything good.
  But my anguish increased;
  my heart grew hot within me.
  While I meditated, the fire burned ...
  You have made my days a mere handbreadth;
  the span of my years is as nothing before you ...
  Hear my prayer, Lord,
  listen to my cry for help;

do not be deaf to my weeping.
I dwell with you as a foreigner,
a stranger, as all my ancestors were.

That was an echo from my own soul, and it called my soul to God. I didn't experience any sudden illumination, but I came back to these words every evening.

Then I read Mark's Gospel as a whole and came to the story of the passion. When I heard Jesus' death-cry: "My God, why have you forsaken me?" I felt growing in me the conviction: there is someone who understands you, who is with you in your crying to God, and who felt the same forsakenness that you are living through now. I began to understand the forsaken Christ, because I knew he understood me. He was the divine brother in need, the companion on the way who goes with you through this "valley of the shadow of death," the fellow sufferer who carries you in your pain. I summoned up the courage to live again and was slowly but surely seized by a great hope for God's "wide space where there is no cramping anymore."

This perception of Jesus did not come suddenly overnight, but it became more and more important for me. I read the story of the passion of Christ again and again and discovered my little life story in his great story.

In the summer of 1946 I realised that my captivity was going to last longer than I had thought. I had been captured late in the war and would be released later than other prisoners. I heard about an education camp in England where so-called "baby-prisoners" could repeat their final school leaving examinations that were required to enter university. This possibility really did exist in the British "camp culture." I applied, passed an English language test and, guarded by a soldier with a rifle, was put on a train and travelled through sunny and so peaceful England to Norton Camp in Cuckney, near Mansfield in Nottinghamshire.

Romantically situated in parkland belonging to the Duke of Portland it was a camp intended to train teachers and pastors for post-war Germany. Set up by the British YMCA, financed by the American businessman John Barwick and guarded by the British army, Norton Camp was England's generous gift to German prisoners of war.

Life for us at this camp was a kind of enclosed monastic existence from which time and the outside world were excluded. The day began at 6:30 a.m. with a bugle reveille and ended at 10:30 p.m. when the lights were put out. Suddenly we had plenty of time and, intellectually completely famished as we were, had set before us a rich library, put together by the YMCA and the prisoners-aid of the WCC in Geneva.

I read everything I could lay my hands on—poems, novels, mathematics and philosophy and any amount of theology, and that literally from morning to night. My first theological book was Reinhold Niebuhr's *The Nature and Destiny of Man*, which impressed me deeply, although I hardly understood one sentence of it. Many international visitors came, among them John Mott, Visser 't Hooft and Martin Niemöller. We had some imprisoned professors of theology who taught imprisoned students theology for free. I have never since lived so intense an intellectual life as I did in Norton Camp. We received what we did not deserve, and lived in a spiritual abundance we had not expected.

A special event which completely turned my life upside down was the first international SCM conference after the war, held at Swanwick, Derbyshire, in Summer 1947. A group of POWs were invited to attend, I being one of them. We arrived with fear and trembling. What were we to say about wartime horrors and mass murder in the concentration camps? But we were welcomed as brothers in Christ and could eat and drink, pray and sing with young Christians from all over the world. To be accepted like that was a wonderful experience.

But then a Dutch student group came and wanted to speak to us officially. They told us that Jesus Christ was the bridge on which they came to meet us and that without Christ they would not have been able to speak to Germans. They told us about the Gestapo terror in their country, about the murder of their Jewish friends, and about the destruction of their homes during the German occupation. We too could step on this bridge which Christ had built from them to us, even if we did so only hesitantly at first, could confess the guilt of our people and ask for forgiveness. At the end we all embraced. I could breathe freely again. And—most important—I knew what I was going to do: Theology.

In April 1948 I went home on one of the last transportations of POWs. I was discharged after five years in barracks, trenches and camps. But I had experienced something that was to determine my whole life. For that reason this time for me was so important that I would not have missed a day of it. What at the beginning had looked like a grim fate became an undeserved blessing. It had begun in the darkness of war and then when I went to Norton Camp the sun had risen. We came with severely wounded souls, and when we went away "my soul was healed."

Jürgen Moltmann: How "Christ the bridge" led me from science to theology

When after fifty years we survivors returned to England in 1995, we visited the site of Norton Camp again. The Nissen huts had disappeared, but the great old oak trees remained. We remembered the people, who that time after the war had met us with forgiveness and reconciliation. And we thanked God with hymns.

The camp's old commanding officer, Major Boughton, said the same thing but more dryly: "I have never before heard of prisoners returning to prison of their own free will and praising God for what they had experienced there." But that was exactly what we did with great joy.

This is for me not only my "personal story," but also the ground and motive for my theological thinking. Relevant theology is always set in life and-death experiences, otherwise it becomes speculative. I experienced hope that enabled me to live—to survive and not give up—in those camps; hope is the last to die, as the proverb says, if at all. These experiences are woven into my *Theology of Hope* (1964). And as I have already told, the godforsakenness Jesus felt between Gethsemane and Golgatha was a consolation to me in my own feeling of forsakenness. This is the foundation of my second book *The Crucified God* (1972), and the reason this book was written in the blood of my heart, so to speak.

I shall turn now to the human, or to be more exact, the personal dimensions of science, because what led me from physics to theology were the human questions of suffering and death, of guilt and forgiveness—in short, the God question.

The question about God and the question of "what holds nature together in its innermost being," as Goethe put it, are not wholly divergent. They are not even controversial. I believe they belong together and whoever separates them is damaging both. According to Plato, wonder is the beginning of all knowledge and truth is beautiful. According to the Jewish-Christian tradition, the fear of God is the beginning of all wisdom. If wonder over the phenomena of nature converges with reverence before the great mystery of the whole world, the outcome is a humble search for truth and unending joy at its discovery.

Knowledge is power

The motto "Wissen ist Macht" was written in the popular scientific journals of my youth to say that science is the method of seizing power over nature. Through science and technology man will become "lord and owner of nature," as René Descartes promised in his book *Discourse on Method*. This promise in fact was fulfilled in the "scientific-technological civilisation" in the nineteenth century. Was this atheistic? No, it was religious, motivated by the Judeo-Christian creation story and explained by modern theology.

Francis Bacon, the author of the phrase "knowledge is power," saw in the gaining of power over nature the redemption of humankind from the fall and liberation from the subsequent dependency on nature. The goal of the scientific elucidation of nature was "the restitution and reinvesting (in great part) of man to the sovereignty and power which he had in his first state of creation." Man was created in the image of God and called to rule over the earth: "Subdue the earth" and "have dominion over the animals."

So, the more man's rule over nature grows, the more he becomes again in the image of God. God is the almighty ruler of the world, his human image must become the mighty ruler of the earth. So knowledge is power and power is divine.

Gaining power was, and is, a modern motive for the exploration of nature, but it is a very dangerous one. The twentieth century saw the politicisation of science as fascist and communist ideologies took possession of the sciences. The twentieth century also saw the militarisation of science during two world wars and the cold war that followed. The twenty-first century has brought us the economisation of science. Knowledge is not only power, but profit as well. These interests are alienating pure science from nature. Do we want to know nature for its own sake, seeking truth and correspondence with nature and finding what we have in common with nature? If so, the sciences are better preserved in fellowship with theology, than in the service of politics, economics and the military.

Has man in the name of God to "subdue the earth" in order to become aware of his own nature as in the "image of God"? One may continue reading the Bible. According to Genesis 1 human beings have to "subdue the earth"; according to Genesis 2 human beings are taken "from the earth," and dying they will "return to the dust." They are not divine lords of the earth, but children of the earth. The earth is not their subject, but their mother. Who can subdue his mother, rob or sell his mother?

In the great dramatic picture of creation in Job 38–40, a human being is very small and insignificant before God's wild and immense creation. "Where were you when I laid the earth's foundation? Can you bind the chains of the Pleiades? Can you loosen Orion's belt?" And Job answers: "I am unworthy—how can I reply?" This is the answer of human wisdom. True knowledge presupposes cosmic humility, as Richard Bauckham maintains, not the "arrogance of power." True science is bound to truth and is not for sale.

Truth is beautiful

I believe an aesthetic dimension is essential for pure science. Where there is truth beauty is also present because beauty is the "splendour of truth." Since ancient times truth and beauty have been seen together: *pulchritudo est splendor veritatis*. "Truth is the truth of being. Where the truth appears the appearance is beautiful," said Martin Heidegger. In the history of science beauty has been taken as a sign that truth is near. Truth has also been recognised in the correspondence between mind and matter, the subject and the object, the *adeaquatio rei et intellectus*. Is this only an old romantic glorification of science, or is beauty still a concomitant of every truth-seeking science?

Boethius in his *Consolation of Philosophy* was formative for the truth concept of classical physics. The beauty of the world is manifest in its mathematical order, because all things find their harmony in numbers. Therefore this world harmony is also recognised in music. Music is the art of bringing different parts to unity through consonance. Biblical wisdom confirmed this worldview: "You ordered everything according to measure, number and weight" (Wis 11:20).

In classical and modern physics the beauty of truth was discovered in the symmetry and order of geometrical figures. Euclidean geometry was for Johannes Kepler "the archetype of beauty." He developed a theory of the *harmonia mundi* out of the discovered symmetry of the constellations of the stars in the firmament.

Another sign of truth is simplicity: *simplex sigillum veri*. The simple is clear and distinct. Next to simplicity unity is a sign of truth. This principle was important for Isaac Newton because the reconciliation of different aspects and the unification of various theories were considered signs of stability. And stability and consistency were considered signs of the faithfulness of the creator of the universe and of the unity of God's creation.

Many modern physicists are on their way to finding the unified world formula. This may be through the unification of different theories, for example the general theory of relativity and quantum physics. One may also try to unify different theories by discovering common symmetries of resonance to each other. Beauty is also at stake in the modern physics of complex systems and in sciences that use morphology. That is the beauty of forms and figures. "Fractal geometry is revealing a world full of beauty," confessed its founder Benoit Mandelbrot. And when we understand the cosmos as an unfinished open-ended process and see the emergence of the new in time, we can discover a "dynamic concept of beauty." The "splendour of truth" is shining not only in timeless symmetry, but also in symmetry-breaking transformation. The becoming and the passing away of temporal forms are beautiful signs of life.

The German scientist Friedrich Cramer, well-known for his *Time Tree* and his *Symphony of Life: Attempting a General Theory of Resonance* stated: "What we call beauty belongs neither to pure order nor to pure chaos. Beauty is revealed where chaos is flowing into order or order into chaos."

What I want to say with these few remarks on the connection between science and aesthetics is this: beauty is still a sign of truth. And this is important for any "truth-seeking community," as John Polkinghorne called the scientific community. Beauty is not a matter of personal taste and "you can't dispute about taste." Beauty is part and parcel of the self-revelation of truth that happens in every discovery. This beauty may be useless, but it is most intrinsically meaningful, which is more than one can say of all technical, political or economic utilisation of scientific discoveries. Scientists may be proud of their joy at the beauty of their discoveries. As Confucius once said:

Better than knowing the truth
is loving the truth.
Better than loving the truth
is rejoicing in truth.

Knowledge is wisdom

Nature is not a "mechanism" (Robert Boyle), nor a "world machinery," nor a cosmos of timeless laws. The nature of the earth is full of wisdom. We humans are part of the earth, and the earth is not a meaningless accumulation of matter and energy, but an ecological system with a long evolution more like an "organism" than a mechanism, more a living object than a dead one. The earth's organism is a subject in its own right. The earth "produces living creatures" (Gen 1: 24). The earth is the only creation God has designed to create living beings. And human beings are taken from the earth.

Jürgen Moltmann: How "Christ the bridge" led me from science to theology

At a certain point in its evolution the earth began to feel, to think, to become conscious of itself and to sense reverence, as Leonardo Boff has said:
And at this point humankind appeared on the scene. We therefore don't stand over against the earth to subdue it; we are, in whatever we recognise or do, part of the earth's community. This cosmic community is wider than the limited parts of nature that we can see and control. Science becomes wisdom wherever we integrate ourselves into the earth's community and into the greater cosmic community.

Wisdom is the ethics of science

Wisdom doesn't come from what we experience but from the way we deal with our experiences. It is not perception by itself that makes us wise; it is the perceiving of perception. It is when we apply our consciences to science and to what we know that the ethics of knowledge emerges. We look back over our shoulder at ourselves, so to speak, and ask "What are you doing? Is it good or evil? Does it serve life or death?" For scientists in the twentieth century the question was "Do we serve peace or war?" We can see the dilemmas involved if we look at two outstanding personalities, Albert Einstein and Fritz Haber.

Einstein's discovery of the general theory of relativity in 1907 was "the happiest thought in my life." Its proof in 1915 through the predicted movement of Mercury convinced him "that nature had spoken to him."

The outbreak of the Second World War in 1939 raised the fateful question: "Develop the atomic bomb in Hitler's Germany or in the Western democracies?" Einstein made his decision and wrote his famous letter to President Roosevelt. The Manhattan Project began and led to the destruction of Hiroshima in August 1945. And with this the Second World War ended, and the nuclear end-time of humankind began. Einstein became a pacifist in consequence: "Abolish war!"

Fritz Haber was a famous German chemist and Nobel Prize winner. His discovery of how to isolate nitrogen from the atmosphere made it possible to produce artificial fertilisers in peacetime, and ammunition in wartime. During the First World War it was Haber's chemical discoveries, that made the German poison-gas war possible. For this Haber was ostracised by the international scientific community after the war.

The commandment of scientific ethics was clear for him: "In peace—for humanity; in war—for the Fatherland!" His love for the Fatherland had however its limits. When the Nazis took over his Fatherland in 1933 he refused to collaborate any further.

How far does the responsibility of scientists for their scientific research go? Does it include responsibility for their results, or even for what others make of their results? In 1957 leading German nuclear physicists recognised this responsibility. In an official declaration they refused to cooperate in the nuclear rearming of Germany, much to the annoyed frustration of leading politicians.

Of course this responsibility is not restricted to the scientists and technicians concerned. It is the responsibility of all the citizens of that society. Scientists share this political responsibility in a special way. And if the whole human race is threatened by a nuclear or an ecological or an economic disaster, the responsibility rests with all human beings. It may be a good idea to discuss the introduction of the Hippocratic Oath for all members of the scientific community again to make clear that we serve life, not death.

Today the formula for power is principally in the hands of the scientists. But a second formula is that for the exercise of ethical power over the physical, and this formula is today still underdeveloped. The last centuries have brought such an enormous increase in scientific knowledge and technological power that what we need in the immediate future is a yet greater increase of wisdom and of wise dealings with what we know and can do. Wisdom is the love of life and the courage to be. Truth is what we seek in science and spirituality. The beauty of God may redeem the world.

> This chapter is the edited text of a Faraday Lecture given at Emmanuel College, Cambridge on 14 February 2012. It first appeared as one article in the journal *Science & Christian Belief* and is used with permission.

34. John Lennox: Christian apologist in an age of doubt

By Chris Mulherin

At a time when Christian thinking is under threat in the West, John Lennox, Oxford professor of mathematics, has become one of Christianity's most passionate and intellectually rigorous apologists. He was speaking at the "Faith has its Reasons" conference on science and faith in August 2014 run by the Peter Corney Training Centre and held at Glen Waverley Anglican Church, Melbourne, Australia.

John Lennox is a comfortably built Oxford mathematician who charms his audience with a grandfatherly demeanour, muted Irish tones and an incisive wit. Lennox is also one of the big names on the Christian apologetics circuit, battling scepticism globally and particularly taking on the so-called New Atheists. He does live shows around the world and there's no shortage in cyber space of Lennox debating such atheist "greats" as Richard Dawkins and the late Christopher Hitchens.

In Melbourne, this time the theme of his conference talks was the relationship between science and Christianity (always pronounced in two parts: Christi—anity) as he spent time carefully explaining many of the misconceptions and contradictions lying behind the idea that there is a conflict between science and Christian faith. There was no lack of anecdotes to illustrate the message, often drawn from his own conversations with sceptics or fundamentalist Christians and rounded out with his repeated refrain (hear the Irish inflection): "So you see ladies and gentlemen ... "
To Christians who insist on interpreting the Genesis record literally, the response is blunt: "Jesus is not a door." Yes, Jesus called himself a door, but he did not mean that literally. The truth lies in the metaphor; Lennox explains that the Bible is full of figurative language which no Christian dreams of taking literally. So he warns about thinking that Genesis 1 and 2 are meant to be day-by-day descriptions of the beginnings of the universe.

Lennox warns about expecting proof in areas of faith and in other endeavours too. "Proof only occurs in pure mathematics," he says, and not in any other discipline; "not in the physical sciences and not in everyday life." But that doesn't mean we can't be sure enough about things to stake our lives on them; we put our trust in aeroplanes and in the love of a spouse.

What we have in such cases is "forensic proof"—proof beyond reasonable doubt.

Commenting on the militant so-called New Atheist movement of the last decade or so, Lennox suggests that Christian thinking is under threat and that cultural forces mean that it is not a level playing field. But, he says, "at least [New Atheist and evolutionary biologist] Richard Dawkins has done this for us: he has put God back on the agenda."

Lennox deals with some of the standard New Atheist one-liners against religion with his own counter thrusts. In response to the physicist Stephen Hawking's claim that religion is a fairy story for people afraid of the dark, Lennox quips that atheism is a fairy story for people afraid of the light.

And Lennox turns the tables on the Freudian argument that religion is a crutch for the weak who project their wishes onto a father figure in the sky: If there is no God, Freud's is a brilliant explanation of why we believe, but on the other hand, if there is a God then atheism is easily explained in similar manner as wish fulfilment for those who do not want to be accountable to their maker.

To the argument that belief in God is like believing in Father Christmas, Lennox responds that while many people become Christians as adults no one starts believing in Father Christmas when they are older. The New Atheists' analogy between God and Santa or the tooth fairy might be good for a laugh but it need not be taken seriously.

As for the extreme claims of some atheists that science is the only road to knowledge—a philosophy known as scientism—Lennox spoke of the differences in types of explanation. An answer in terms of physics to "Why is the water boiling?" is adequate at one level. But it leaves out essential elements. "Because I want a cup of tea," is also part of the fuller account for the boiling water. So it is with the universe and the human race; while science can go so far, it cannot offer answers to the questions of meaning and purpose; "God and science are not logical alternatives, they are logically complementary."

Not only is there no conflict between science and Christianity—evidenced by believing scientists over the ages—but, says Lennox, science arose in part because of the Christian worldview that understood nature to be governed by the laws of a rational God. Lennox cited Peter Harrison, the Australian historian of science, recently returned from Oxford, who argues that monotheism and then Christianity had an enormous effect on driving science. "We owe science in the modern sense to Christianity," he says, but "science has now outgrown its cradle."

Lennox also reiterated the age-old claim that atheism is unable to offer a foundation for objective morality. The traditional base for ethics is transcendent, he said, citing the 10 commandments, for example, enshrined in the laws of most Western countries. So the question of a foundation for ethics or a moral compass is back on the agenda. In fact, the New Atheism's "success" has resulted in other much more philosophically serious atheists warning of the dangers of the exclusion of religion from the marketplace of ideas. Lennox quoted the highly respected social theorist Jürgen Habermas who is concerned about "an unfair exclusion of religions from the public sphere" and the danger of "severing secular society from important resources of meaning."

Lennox says that one of the challenges for Christian communication is the confusion about the nature of God. Christians can't assume people understand what they mean by God. Many people think that God is identical to an ancient Greek god—a "god of the gaps," which is invoked to explain phenomena such as lightning. But, he says, if you define God as a god of the gaps then you have to choose between God and science, which progressively fills the explanatory gaps in our understanding of nature. The Christian God is not of that order; Christians need to explain that God is the creator and sustainer of the universe and of the laws of nature themselves.

While the gods of the Ancient Near East descended from the heavens and the earth, the God of the Bible made the heavens and the earth; the God of the Bible is not of the gaps but of the whole show, and the more we understand of the universe, the more we can admire the God who made it.

Lennox was also keen to emphasise similarities between science and faith; Christianity, like science, is based on trust and on evidence. He was quick to point out that there are different types of evidence but that biblical faith and science share this in common: they are both based on commitment to truth, and believing according to the evidence. Lennox quotes Einstein who said that he couldn't imagine a scientist without faith in the rational intelligibility of the universe. And Anglican priest and physicist John Polkinghorne says that physics is powerless to explain its faith in the intelligibility of the universe; such a belief is taken up by the scientist before starting to do science.

Lennox also raised "Darwin's doubt," the question of whether non-theistic evolution gives grounds for believing that humans can grasp truth. Charles Darwin put it this way in a letter written 20 years after his *Origin of Species*:

The Human Story

With me the horrid doubt always arises whether the convictions of man's mind, which has been developed from the mind of the lower animals, are of any value or at all trustworthy. Would anyone trust in the convictions of a monkey's mind, if there are any convictions in such a mind?

35. Antony Flew: The philosopher who flew the atheists' nest

By John Pilbrow

Antony Flew's courageous abandonment of atheism in the last six years of his life demanded much of him, as it does of us. John Pilbrow reflects on one philosopher's pilgrimage from unbelief. (Numbers in brackets refer to pages in Flew's *There Is a God.*)

At the beginning of a public debate at New York University in May 2004, one of the world's most notable atheist philosophers, Antony Flew (1923–2010), announced "that he now accepted the existence of a God" (p. 74), something that sent shock waves through the atheist world! But it showed that atheists sometimes do change their minds. Given the triumphalism exhibited by some atheists, for example, during the Global Atheist Convention in Melbourne, Australia in 2012, a reflection on Flew's defection from atheism is worth undertaking.

In 2007, Flew traced his journey from adolescent atheism and Marxism to Deism in a little book, *There Is a God: How the World's Most Notorious Atheist Changed His Mind*. Here we learn why he found it necessary to abandon many of the views he had previously expressed in support of atheism. Indeed, he went so far as to describe his book, *God and Philosophy*, as an historical relic! He admits that he embraced atheism too quickly in the light of the problem of evil. This is perhaps understandable given what he experienced during family visits to pre-World War II Nazi Germany as a teenager.

Although Flew departed from the Methodism of his childhood, he remembered his theologian father's advice to follow the path of wisdom and also Plato's dictum "to follow the argument wherever it leads." After his conversion to Deism (the idea that there is an impersonal mind behind the universe), he mused, "though I had come to see things from a different perspective, I was still playing the game with very much the same passion and principle as before" (p. 65).

Flew's reputation as an atheist apologist was established by his short, but very influential and widely disseminated essay, *Theology and Falsification* (1950), first presented as a paper to the Oxford Socratic Society. His objectives were "to clarify the nature of the claims made by religious believers" (p. 43) and "to spice up the bland dialogue between logical positivism and

Christian religion" (p. 44). Further, he recognised that the "presumption of atheism is, at best, a methodological starting point, not an ontological conclusion" (p. 56). (The preface to *There Is a God* by Roy Abraham Varghese gives a very readable overview of Flew's thought over six decades).

While Flew recognised that science on its own "cannot furnish an argument for God's existence" (p. 155), he long believed that the onus of proof lay "on the one who affirms, not on one who denies." Therefore, he used to believe that it was up to theists to prove God's existence (p. 69). In his later years, Flew was strongly influenced by several Christian philosophers, in particular, American Alvin Plantinga and his assertion that there exist basic beliefs such as theism and God's existence (p. 55). This was an "eye-opener" for Flew who eventually realised that such basic beliefs were not derivable from within philosophy itself. In due course he concluded that the burden of proof actually rests with atheists themselves (p. 56): "Nevertheless, if Plantinga is right, and I think he is, much of the discussion regarding proofs for the existence of God becomes irrelevant."

Flew was also critical of Richard Dawkins' claim that God, if he exists, must be complex and he advanced arguments that God must, in fact, be simple. And he rightly pointed out that the old atheists would never have claimed God was a scientific hypothesis!

Flew's journey and his eventual change of mind, at least to Deism, was also strongly influenced by modern science. Nature obeys laws, intelligent life exists and nature exists (pp. 88–89). Flew stated:

> Those scientists who point to the mind of God, do not merely advance a series of arguments ... Rather they propound a vision of reality that emerges from the conceptual heart of modern science and imposes itself on the rational mind. It is a vision that I find compelling and irrefutable. (p. 112)

He identified four important issues for further investigation: the origin of the laws of nature, how life originated from non-life, how the universe came into existence and "Did something come from nothing?" (p. 91). Flew refers to the zero energy, quantum fluctuation idea that might explain how the universe arose from nothing (p. 142). However, at present, physics can only look back close to, but not exactly, to time zero 13.7 billion years ago. It is silent about the "before" and speculation about what went "before" is in any case metaphysical, not scientific.

There is a particular matter discussed during Flew's debate at New York University with Jewish physicist/theologian Gerald Schroeder and Scottish

Antony Flew: The philosopher who flew the atheists' nest

Christian philosopher John Haldane, already referred to at the beginning of this chapter, that needs to be addressed. There Schroeder spoke about an experiment in which six monkeys and a computer were placed in a cage. At the end of one month, no single word in the English language had been typed. Schroeder then demonstrated the improbability that six such monkeys could produce a Shakespearean sonnet by chance. While Flew found this line of argument incredibly persuasive, I disagree strongly with his conclusion: "If the theorem won't work for a single sonnet, then of course it's simply absurd to suggest that the more elaborate feat of the origin of life could have been achieved by chance (alone)" (p. 78). In any case, Schroeder's argument doesn't really help. Why not? Because we can't be sure about what is or is not possible without evidence. The reality is that life exists (with a 100 percent probability) and modern science has furnished us with robust levels of understanding concerning cosmic history and the evolution of life during the past 13.7 billion years. Flew's mistake was to assume that evolution operates by chance alone, but this is not so (see J. C. Polkinghorne, *One World: The Interaction of Science and Theology*, SPCK, 1986, pp. 50–51).

In a discussion of molecular biology, complete with quotes from high-profile biologists who profess ignorance about how life might have arisen from non-life, Flew concludes that "the only satisfactory explanation for the origin of such 'end-directed, self-replicating' life as we see on earth is an infinitely intelligent mind" (p. 132). It is one thing to claim that the emergence of life was the result of an intelligent mind (creator), with which I agree, but it is quite another to imply that the step from non-life to life required a special intervention, rather than through processes embedded in the natural order by the creator. It seems to me that Flew confused God (as creator) with processes of cosmic and biological evolution.

Nevertheless, Flew showed considerable courage in retracting much of his previous support for atheism and it is a tribute to his intellectual integrity. We should be grateful that he exposed many weaknesses in arguments used by the New Atheists (see Varghese's critique of the New Atheism in Appendix A of Flew's book). Rather than feeling triumphant, the challenge for Christians is to be clear about what they believe and why.

During the three years that elapsed between publication of *There Is a God* in 2007 and his death in 2010, it is reasonable to wonder whether Flew moved nearer to the heart of Christian faith. But he did say "no other religion enjoys anything like the combination of a charismatic figure like

Jesus and a first-class intellectual like St Paul" (p. 157). Appendix B, by the former Bishop of Durham, Tom Wright, finishes the book with a historical evaluation of the gospels and the life, death and resurrection of Jesus, that goes beyond the philosophical arguments found throughout the book.

36. Francis Collins: Work on human genome strengthened his faith

By John Pilbrow

Francis Collins is a brilliant scientist and former atheist, for whom the deciphering of the human genome played a significant part in the deepening of his faith.

What is it about the lanky North American, Francis Collins, with a passion for riding motor bikes, composing songs and playing the guitar that warrants a chapter in this book? *The New Republic* magazine of November 2011 considered him the third most powerful, least famous person in Washington! Why? As Director of the US National Institutes of Health (NIH; annual budget > US$30b) since 2009, he evokes more than passing interest because he is also an evangelical Christian who has participated in debates with atheists such as Richard Dawkins. From 1993-2008, as Director of the US Human Genome Research Institute, he also had the oversight of the International Human Genome Sequencing Consortium, one of the most significant international scientific projects ever undertaken (usually referred to as the Human Genome Project, HGP).

Collins is an eminent medical geneticist whose reputation was secured through contributing to development of chromosome jumping, a method that significantly speeded up searching for specific genes. This method enabled Collins and his group at the University of Michigan along with Canadian collaborators to locate genes responsible for crippling diseases such as muscular dystrophy and Huntingdon's disease.

In 1998, Celera, a private company, headed by geneticist Craig Venter, announced plans to sequence the entire human genome independently providing a real test of Collins' leadership of the HGP. Two years later, on 26 June 2000, in the East Room of the White House in the presence of President Bill Clinton, Collins and Venter announced that they had assembled a joint working draft of the entire human genome consisting of some three billion base pairs. The final version was released in 2003.

When President Obama nominated Collins as Director of the NIH, several politicians, certain prominent academics and some in the media expressed concern that Collins, as an evangelical Christian, might have wanted to maintain the Bush era restrictions on stem cell research. It appears that sensible boundaries for future research are now in place and Collins has handled controversies such as these with commendable forbearance and Christian grace.

Given his high public profile and Christian convictions, Collins' book, *The Language of God: A Scientist Presents Evidence for Belief* (Free Press, 2006), not surprisingly has evoked a range of responses. Written in an easy-to-read style, it includes a birds-eye view of modern science, ranging from cosmology to the deciphering of the entire human genome ("God's instruction book").

Raised on a North Carolina farm, Collins was home schooled until the age of 12. While doing his PhD in physical chemistry at Yale, and aware that science could explain so much, there seemed to him, as an atheist at the time, no room for what he judged to be unthinking faith! A course of lectures on biochemistry opened his eyes to the living world and led to his decision to study medicine. On being confronted by dying patients, Collins was challenged to explore some of life's big questions. Turning to *Mere Christianity* by C. S. Lewis, he began to find some of the answers. Like Lewis, his journey from atheism to a personal faith in Christ did not happen overnight.

In *The Language of God*, Collins disputes atheist claims that science, and in particular, evolution, inevitably leads to atheism, contending that science is not the only way of knowing. Further, he refutes creationist arguments that suggest faith is opposed to science, and also exposes weaknesses in arguments for intelligent design.

An unashamed evolutionist, Collins contends that,
> no serious biologist today doubts the theory of evolution to explain the marvellous complexity and diversity of life. In fact, the relatedness of all species through the mechanism of evolution is such a profound foundation for the understanding of all biology that it is difficult to imagine how one would study life without it. Yet what area of scientific enquiry has generated more friction with religious perspectives than Darwin's revolutionary insight? ... through to today's debates in the United States about the teaching of evolution in the schools, this battle shows no signs of ending. (p. 99)

It turns out that only about two percent of the human genome is involved in making us what we are. The remaining 98 percent, thought to be inactive, includes ancient repetitive elements (ARE's) about which Collins comments:

Unless one is willing to take the position that God has placed these decapitated AREs in these precise positions to confuse and mislead us, the conclusion of a common ancestor for humans and mice is virtually inescapable. This kind of recent genome data thus presents an overwhelming challenge to those who hold to the idea that all species were created *ex nihilo*. (pp. 136–7)

In spite of the evidence in support of evolution, Collins is deeply concerned about the impact of young earth creationism on young people, saying,

[it] does damage to faith, by demanding that belief in God requires assent to fundamentally flawed claims about the natural world. Young people brought up in homes and churches that insist on creationism sooner or later encounter the overwhelming scientific evidence in favour of an ancient universe and the relatedness of all living things through the process of evolution and natural selection. What a terrible and unnecessary choice they then face. To adhere to the faith of their childhood, they are required to reject a broad and rigorous body of scientific data, effectively committing intellectual suicide. Presented with no other alternative than creationism, is it any wonder that many of these young people turn away from the faith, concluding that they simply cannot believe in a God who would ask them to reject what science has so compellingly taught us about the natural world. (pp. 177–9)

As much of the science–faith debate in America is highly polarised, Collins founded BioLogos (*logos* = Word, as in John 1.1), an organisation devoted to raising the level of the science–faith conversation (see biologos. org). He coined BioLogos as a way of affirming that God has endowed the living world with the capacity for diversity and ultimately intelligent life.

The basis of the harmony Collins seeks and also finds, rests on the essential unity of truth. For those who believe in God, Collins argues "there are reasons now to be more in awe, not less" (p. 107). In so saying, he joins a large chorus of thoughtful scientists, theologians and philosophers such as Alister McGrath and John Polkinghorne.

Collins' reflection on the HGP is particularly revealing:

For me, as a believer, the uncovering of the human genome sequence held additional significance. This book was written in the DNA language by which God spoke life into being. I felt an overwhelming sense of awe in surveying this most significant of all biological texts.

Yes, it is written in a language we understand very poorly, and it will take decades, if not centuries, to understand its instructions, but we had crossed a one-way bridge into profound new territory. (p. 123)

The overriding theme of *The Language of God* is well summed up in the final four sentences:

Science is not threatened by God; it is enhanced. God is most certainly not threatened by science. He made it all possible. So let us together seek to reclaim the solid ground of an intellectually and spiritually satisfying synthesis of all great truths. (pp. 233–4)

I share Collins' concern as to why so many people seem upset at the idea that we have evolved. Since God entered the world in the person of Jesus Christ, our dignity is assured. More important than how we got here is the fact that that God reaches out to us.

37. Interviewing Alister McGrath: Science is helpful, but only part of the picture

By Chris Mulherin

The Rev. Dr Alister McGrath, an Anglican priest and former atheist, came to faith at university, where he discovered that Christianity could be both life-giving and intellectually exciting. Professor of Science and Religion at Oxford University, Director of the Ian Ramsey Centre for Science and Religion, and Professor of Divinity at London's Gresham College, he believes that science and religion offer differing and enriching perspectives on how to live a meaningful life.

Atheism, Christianity and the journey of faith

Chris Mulherin: Alister, you talk often about your atheism when you were in your teens and then coming to Christian faith. I wonder if you could just give us a potted history?

Alister McGrath: Well, I was an atheist as a teenager at school. I was studying sciences and it did seem to me that if I studied science and loved science that I couldn't in any way be involved in religion, and I became quite an aggressive atheist. I took the view that science completely eroded the conceptual space occupied by God and I could not really see why anyone would want to be religious.

So when I went up to Oxford in 1971, I was pretty clear that I was expecting my atheism to be confirmed and I had no sense that in fact it would be radically undermined and redirected. I think what really did it was a growing realisation, partly through talking to other atheists and partly talking to Christians at Oxford, that atheism was radically evidentially underdetermined.

In other words, it might make some sort of sense of life, it might be internally consistent as a worldview, but actually the evidence for it wasn't that good and it did not deliver existential benefits. It was existentially bleak and austere. Whereas Christianity was intellectually exciting and that was something I had not realised before. That was the gateway to the discovery that's characterised the rest of my life, which is that Christianity is *generative*: it opens up ways of thinking, ways of conceptualising the world, which don't just help us to make sense of it but help us to cope with life and actually to take delight in this created order, knowing that one day something better will be at the end of that journey for us.

The gateway experience of wonder

Part of the journey that you've described is the place of wonder, meaning, value, purpose ... What's the place of wonder in your journey?

Well, it's played a very big role for me. I think I experienced it a little bit when I was about nine or ten years old. When I began to get interested in astronomy, I began to realise how big the universe was—not just mathematically how big it was, but actually looking at the night sky and being overwhelmed by its beauty and its vastness.

I think, for most people, the experience of wonder is generative. What I mean by that is it makes you ask other questions: "Is there anything behind this or is this it?" And that's one of the reasons why I've so much enjoyed engaging in debate with Richard Dawkins, because Dawkins says, yes there is this sense of wonder but that is simply it; there's nothing beyond. Whereas for me, the full significance of wonder is not just this sense of delight at the existing order, but this transcendent realisation that this is only the beginning.

Do you know the image that Albert Einstein uses? He says, "I sometimes think we are just holding the lion by its tail." And that's a very powerful metaphor. We've got hold of something but it's only part of something much, much bigger that we need to discover. And so, for me, wonder is a gateway experience. It opens doors; it expands your mind and makes you realise that there is very likely more to life than we realise, and wonder what there is that remains to be discovered.

A moral universe?

You mentioned Richard Dawkins and his sense of wonder, but he doesn't think that anything lies behind it. He also has a very strong sense of morality yet he says we live in a universe that has no good and it has no evil. Is that a contradiction in Dawkinsian thinking?

I think some people would say that there is a clear contradiction between Dawkins saying there is no purpose, no meaning, no good, no evil, and his own adoption of certain moral values. Personally I think that he illustrates very well what happens if you do say there is nothing good or evil embedded in the universe that we need to respond to, and therefore we make it up. So

Interviewing Alister McGrath: Science is helpful, but only part of the picture

we say, "Well I think this is right, this is wrong." Good is what I choose it to be and if there is nothing beyond me to say "I'm sorry that's not right," then I am ultimately the arbiter of good and evil. So I invent my moral system and I remain obedient to it but it's an act of self-service. It's just me living out the morality I choose and I've chosen to embrace.

So I think you could say Dawkins is offering a consistent viewpoint, but the real difficulty is that there is this massive question, "Why did you choose that moral system and not something else?" and that of course opens up this whole question about whether ethics is just something we invent or which our power group or in-group invents, or whether actually there is something deeper embedded within the structure of the world that we are trying to respond to.

So in one sense it depends on how we look at the world, doesn't it? I think in one place you talk about the natural world being conceptually malleable and in another place you talk about the different perspectives, for example, that we can have about Mount Ben Nevis, the highest mountain in Britain. How does that relate to a view that says there is one truth rather than there are many truths? I guess we're talking about relativism ...

We are, yes. Let's go back to Richard Dawkins for a moment. For me, one of the most interesting things about Dawkins is this: he studies evolution, he knows the mechanisms very, very well. And then he says we must resist this whole thing. In other words, "This is natural. Darwinism is natural. It's all about the triumph of the strong, but we don't want that in our own societies. We want something different." His argument is, "Look, this is what's going on in the biological realm but it doesn't apply to human beings. We need to do something better than this."

The really interesting question is, where does Dawkins get his perception that there has to be something better than this? On the basis of what, is he able to say that Darwinism is not appropriate for human beings? What moral criteria can he use to actually make that judgement? That is why it's so important to talk about multiple perspectives on reality. Because one of the things I would say is that a scientific perspective on life is really very, very helpful but only fills in part of the picture of life.

Science and religion: a stereoscopic view

I think we need to talk a little bit more about that relationship between science and faith. There are various views of that relationship: some see them, as the "New Atheists" certainly do, as a direct conflict of some sort or other, and many Christians, perhaps fundamentalist Christians, would agree. You wrote in Inventing the Universe that you don't wish to merge science and religion, blurring their boundaries. How do you describe the relationship?

If you like, there's a big picture; science fills in some of it, faith fills in other bits of it, but you need both of those sections filled in to lead a meaningful life. Thinking about multiple perspectives on reality is not in any way invalidating a scientific perspective; it's not invalidating a faith perspective. It's saying they illuminate different aspects of reality and we do need them both. We need this stereoscopic vision, if I can put it like that. And Dawkins is offering us a kind of one-eyed vision on life, which lacks existential depth. For me, seeing science and religion as offering differing perspectives, but potentially enriching perspectives, actually gives you this framework for holding science and faith together, but also for living out a meaningful life in this world.

To use an analogy, think of a Venn diagram. There is some overlap but actually there's an awful lot of separate space and one of the key points is that where there is overlap, we have to ask whether there is competition or potential enrichment. One of the real difficulties here is the claim for absolute exclusiveness which unfortunately is true for certain kinds of scientists and certain kinds of religious people: "We have the full picture, anything else claiming to tell the truth is a threat and we need to neutralise that." And the easiest way of doing that is simply by declaring it—if you're a New Atheist—to be irrational and superstitious, or—if you are a religious person—as heresy or unbelief or godlessness or something like that. And really what I'm saying is that where there is overlap we need to do some reflection as to whether we need to make a choice or whether we need to realise there may be limits on both science and faith. And one of the points I try to emphasise is that the fact that science is provisional is a very important element in this discussion.

Provisional?

In other words, scientists can say, "At the moment we think the best answer to this question is this," but then as a footnote, "in 20, 30, 40 years' time we may have changed our minds completely on this." I think one of the real difficulties in the science and religion dialogue is that people have said that science tells us what is right, period. And of course, the real difficulty is that science moves on. I mean, if we were having this conversation a hundred years ago, we would believe in the eternity of the universe: the universe has always been here. We don't believe that any more. It's a massive change and science is like that.

So, one of the things we need to say about the science–religion dialogue is that it's best done by people who are prepared to be modest intellectually and say "I think I've got a very good handle on this, but nevertheless there are points where things are a little bit opaque where I'm not absolutely sure, but nevertheless it looks to me like this."

That's why I personally like the dialogues in science and religion because it forces me to be intellectually humble. In other words, realising—particularly in theology—we haven't quite got this as clear as we would like or sometimes we take wrong turns. That's one of the reasons why that great slogan about the Reformation, "Ecclesia semper reformanda," is so important: the church always reforms itself by keeping asking itself: "Have we really got this right or we do need to come back to this and revisit it?"

The limits of knowledge in science and theology

Discussing the provisional nature of science in the way you do seems to limit the ability of science to grasp truth. How do we understand the claim of science, or the claims of other people like Christians, to have grasped truth, to make truth claims, if they're not in some sense proven or objective truth claims?

I think the real issue here is the human situation: we have a limited grasp on reality and our language is simply not adequate to describing or talking about God properly. Nor is it really adequate to describing the universe. We struggle, whether you're a scientist or a theologian, to use words meaningfully to refer to greater realities that lie beyond us. So actually science and theology are in very similar situations here.

The difficulty that both science and theology face is that we come to think that our present descriptions of things actually are the way things really are. In terms of science, it fails to take account of the fact that science is on a journey and as it discovers more and more evidence, and develops more sophisticated theoretical reflections, it's going to change its mind on certain things. The problem is we cannot predict the direction of that trajectory. In other words, we can do the sort of thing that Michael Polanyi does and say, "Well, as a scientist, I believe all these things to be true but I know some will be shown to be wrong and unfortunately I don't know which of them they are." And Polanyi's point is we need to have personal commitment to these beliefs while at the same time realising that they could be shown to be wrong.

What about theology?

In the case of theology I think it's slightly different. The issue here is that we have real problems with our language. If we try to say something about God, the difficulty is that sometimes we assume that because we say "this," it is exhaustive, that it somehow adequately says everything. Often when we say something about God—"God is trustworthy"—we need to realise there's a lot more that needs to be said.

So both science and theology need to come to terms with the fact that we struggle as human beings to take in this greater reality that lies beyond us. Inevitably that means that we tend to reduce reality to something that we find to be intellectually manageable, and that sometimes leads us to reduce God to the realm of the rational, or in the case of science, to simply use common sense to resolve debates that actually are much more complex. You might think of the very common-sensical reaction to quantum mechanics—"Now how on earth can things be both particles and waves?"—and you just have to say, sorry that's just the way it is, even though it causes severe mental difficulties for us.

Some people who expect science to offer us fairly clear-cut and proven truth, would say that climate change is an example where scientists can't offer proof and, in fact, we're all being duped into a sort of groupthink to believe that human beings are changing the climate.

I think there are many debates in our culture where, in effect, there are many options for scientists. I think, for example, of the big debate about whether there is a multiverse or whether there's just a single universe. When the evidence is open to multiple interpretations the difficulty for scientists is that very often they find themselves having to make a decision. The evidence itself is not conclusive and the intellectually honest thing to say is, "Look I do not think we can reach a safe decision on the basis of the evidence available, but nevertheless some provisional decision does need to be made."

The scientist as a person of faith

In describing science this way, are we saying, in effect, that the scientist too, even the atheist scientist, is a person of faith?

Every scientist I know is a person of faith because they will say "Look I can't actually prove this to be right but I think it is right." The hysterical aversion of some scientists to that word *faith* is more an indication of the rhetoric of the New Atheism than anything else. Absolute proof is limited to the very restricted domains of logic and mathematics.

That's a very disconcerting view, isn't it? And it's certainly not the view of most people in the street about science. What do you think we—we as the Christian church, if you like—ought to be doing in terms of explaining these sorts of things in the public marketplace?

I think there are several things. One of them is to realise the critical role that Christians who are scientists are able to play in advancing public discussion of this. There's a lot of empirical evidence that suggests that the most credible people to talk about the relationship of science and faith are people who have personally integrated this. In other words you can say, "Well, Richard Dawkins says this, but, I'm a scientist, I'm a Christian, here's the way it looks to me." They have a credibility, an authenticity which actually helps people to see how these things can be held together.

Fine tuning, design and apologetics

I wonder if we might talk about apologetics—the art of offering a reasoned defence of the faith? For example, the idea that the universe has certain characteristics that seem fine-tuned for life. Do you think that looking at the cosmos gives us clues about the creator?

Well, fine-tuning is the observation that the fundamental constants of the universe, which could have been very different seem to be set to preassigned values that are life-friendly. The question is, what does that show, if anything? I would say it *proves* nothing. But that's not the point; the key point is you can't prove anything in this area. What needs to be said is that this is absolutely consistent with the Christian understanding of creation.

In other words, if you take the Christian worldview and look at the world, there's a strong degree of resonance between the theory and the observation.

Richard Dawkins would say, "Well it's just a happy accident. Why are you so excited by this?" Of course the key here is that Dawkins is looking at this phenomenon through his lens and I'm looking at it through my lens. We see different things and it brings home the question: Which of these ways of looking at the world is the best?

That brings me back to a very famous point made by G. K. Chesterton who says, "With Christianity it's not this argument or that argument; it is the fact that the whole thing, when you begin to look at its correspondence between what we observe and what we experience, it actually works out." And this is a classic example of this: this isolated observation of fine-tuning on its own proves nothing but it's part of a much bigger context and it has a cumulative force along with many other things which point in the direction of a Christian way of thinking.

Christianity: The best explanation?

Is that akin to saying that Christianity is the best explanation of what we take to be the givens of the world?

In science we look at the question of which of multiple explanations might be the best, realising that we cannot prove that this is the best explanation but nevertheless we can begin to articulate criteria which helps see why this explanation is to be preferred to that.

And in the case of things that go beyond the capacity of experiment to resolve, for example, the question of whether there is a God, then what we need to try to do is deploy as many criteria as possible. For most Christians, I think the answer is something like this: "I don't believe in God because of looking at the world. It's much more I believe in God, and when I look at the world through the God lens I'm just astonished by how good the fit is." It goes back to John Henry Newman who once said, "I don't see design in the world and believe in God; rather I believe in God and as a result I see design in the world."

What Christians need to realise is that their faith gives them this remarkable conceptual lens which allows them to look at the natural world, appreciate its beauty to a far greater extent than would otherwise be the case, and also appreciate the intellectual capaciousness of the Christian way of thinking which helps make so much sense of what we experience within us and what we observe without us.

Apologetics: defending the faith

You're an apologist; you have spent much of your life defending the Christian faith in the public sphere ... what do we need to do in terms of engaging publicly and defending the Christian faith?

I think that many Christians feel disincentivised from doing apologetics; they feel that they will not be able to respond to the questions that their friends and culture at large are asking about faith. When they hear their friends say, "How can you believe in God when there's so much suffering in the world?" they interpret that as a critique of faith. They need to see it as an opportunity to talk about faith and say, "That's a very interesting question. Let me try and explain what Christians think about this."

We need to be much more confident about our faith, in the right sense of that word and feel we can give answers to those good questions that our culture is asking. So, I'm really asking for a change in the mindset of the churches. Clergy and Christian leaders have such an important role to play: partly in dealing with apologetic issues in their sermons, partly in pointing to good apologetic resources, and also in terms of mentoring people and saying, "Look, maybe you can talk about how your faith affects you in this area and that would be really helpful to other people."

At every level of theological education we need to help those who are going to be Christian leaders to help their congregations to give good answers to the questions that their friends are asking. In England, everyone reads C. S. Lewis. And one of the reasons is that they aren't getting this kind of stuff from the pulpit. If we had more apologetic material coming from our pulpits, things might change. Tim Keller, in New York, is a very good example of someone who does a lot of apologetic work in his sermons; there's a model there.

A final question ... what advice would you give to a young student studying science for the first time ... say they were not a religious person but they were interested in the big questions?

That sounds like me when I was 18. Here's what I would say: Take delight in the beauty and complexity of this world in which we find ourselves and enjoy what you see. But always be alert to this: maybe what you are seeing is a pointer to something even better. Just allow that experience of wonder to open your mind to something deeper that might lie behind and beyond what we do see.

38. Allan Day: A scientist's quest for what it means to be human

By John Pilbrow, Alan Gijsbers and Denise Cooper-Clarke

Emeritus Professor Allan Day, Professor of Physiology at the University of Melbourne from 1967–1998, was a leading figure in the dialogue between science and faith for more than 50 years. Here, various aspects of Allan Day's thinking at the faith–science interface are discussed.

Allan Day and the faith–science interface—John Pilbrow

I was privileged to know Allan for some 45 years as a friend and also as a Fellow of both the RSCF (Research Scientists' Christian Fellowship) from 1968 to 1978, and ISCAST—Christians in Science and Technology from 1990.

Allan was convinced that God is the author of all truth, and that the development of modern science presupposed a God of order who created a world that was contingent on his will and could be explored by experiment, not just discovered by reason alone.

Over many years, Allan, accompanied by his wife, Joan, regularly attended conferences of the Ian Ramsey Centre in Oxford and Christians in Science (UK). These visits helped cement friendships with John Polkinghorne, Arthur Peacocke and Oxford nuclear physicist and Catholic layman, Peter Hodgson.

Allan Day was highly qualified with Msc, MBBS and MD degrees from Adelaide, DPhil from Oxford and DSc from Melbourne. In retirement he added a BTh.

Ridley–ISCAST Lectures on science and faith

Allan was justly proud of the 10-lecture ISCAST/Ridley College course, presented in 1997, 1999 and 2001, that he developed with inputs from geologist, Jonathan Clarke, physicist, John Pilbrow, and agricultural engineer, Ross Macmillan. This is what he had to say about it:

> Over the last decade or so an increasing interest in the relationship between science and religion has developed … Oxford University … appointed its first Professor of Science and Religion within the Faculty of Theology in 1999.

The course covered the historical background to the development of science and the interaction of such development with religious faith, philosophical issues with respect to religious and scientific knowledge and consideration of various positions held with respect to the science–faith interface. The approach to science and faith is presented in a way that is consistent with both scientific integrity and biblical authority. An updated version of the course is available at http://www.ISCAST.org/resources/notes_on_science_and_christian_belief

Critique of Intelligent Design (ID)

In the light of the short-lived publicity concerning ID, and a DVD that was widely circulated at the time, Allan wrote a well-informed letter to the editor of *The Age* in August 2005 followed by a most insightful critique in *The Melbourne Anglican* (September 2005). An edited version may be found at http://www.ISCAST.org/journal/opinion/Day_A_2005-11_Intelligent_Design He concluded that ID is bad science, bad metaphysics and bad theology and represents a return to the idea of the God of the gaps.

Allan, anthropology and the human person—Alan Gijsbers

As a second and third year medical student at the University of Melbourne I sat through many physiology lectures from Prof. Day. His mind worked much faster than his speech. As soon as he made a point he would modify that point with a caveat or two. These caveats would have further caveats!

My own debt to Allan is that he introduced me to ISCAST and encouraged me to explore Christian anthropology in relation to addiction and neuroscience. This has been a most fruitful collaboration and I am deeply indebted to him.

Day showed how to take the Bible seriously, as Scripture, and science seriously as a search for truth. Both are needed. The Bible is not a science textbook but is revealed truth in terms of everyday culture and language. In his major scholarly work *Adam, Anthropology and the Genesis Record: Taking Genesis Seriously in the Light of Modern Science*, Day claimed:

> [Genesis 1] is not in any way in conflict with modern science, but brings important principles that themselves form a context for science, giving nature purpose, and humanity a responsibility to God as creator.

Thus the first creation story, Genesis 1:1–2:4a, is about creation by the word of God. Though based in history, it could be called myth, saga or proto-history yet it "conveys primary theological truths."

The second creation story of Genesis 2 and Adam in Eden of Genesis 3 were considered in the light of human pre-history and human evolution. *Homo sapiens* first appeared in East Africa about 200,000 years ago. More sophisticated art and religion appeared 15–30,000 years ago and language some 30–35,000 years ago. Today, genetic data provide a molecular clock that traces mutations back to locate our common ancestors. For example, humans share 98 percent of their DNA with higher apes.

Day recognised that the Hebrew term translated Adam can be generic for humankind or refer to a specific individual. While there is ambiguity about when exactly Adam is identified as a particular human being, who was he? Development of the image of God could have occurred either gradually or instantaneously and the fall could be an event or a process. He then asked whether the New Testament's account of Adam as a historic being is binding on Christians today.

Finally, he posed two questions: Who are we as human beings, and how is that presented in Scripture? He concluded that we have biological and theological dimensions combined in one person, created by God for God, and that in Scripture all of humanity are "in Adam," made in God's image, and stewards of God's creation.

Reflections on the legacy of Allan Day in bioethics—Denise Cooper-Clarke

Waiting for my first class at Ridley College (in theology) I was feeling nervous, all my previous study having been in science and medicine. My nervousness increased when I saw my old grade six teacher, Mrs Bryant, and my former physiology lecturer, Allan Day, both then retired. I wouldn't be the oldest student in the class but the academic bar had just been raised.

Allan Day was ahead of his time and not afraid to go against the views of many conservative evangelical scholars in the area of bioethics. This surprised me in someone of his age who had been a Baptist! After joining St Hilary's Anglican Church, he told me that he was too radical for the Baptists now.

Allan Day produced four important articles on medical ethics:
"Using the Bible as a Basis for Medical Ethics" (*Luke's Journal*, 2001);

"What Does it Mean to Be Human? Theological Responses to Contemporary Biology" (*The Melbourne Anglican* 2002); "Biotechnology and Medical Ethics: Thinking Biblically About Contemporary Medicine" (ISCAST *Online Journal*, April 2002); and, "Why All the Fuss About Stem Cells?" (*The Melbourne Anglican*, 2006). The arguments presented in these papers were based on a number of common foundations. For example, acknowledging that "the Bible knows little of contemporary technological medicine," Allan asked, "How then do we as Christians, who accept the Bible as a guide in matters of faith and conduct, use it as a guide for medical conduct and ethics?" He cautioned against superficial application of biblical texts, arguing instead for the "more onerous task of proper exegesis" and the enunciation of principles relevant to our own context. He asserted, rather boldly, that issues such as abortion, IVF technology and embryonic stem cells, were not "as clear cut as might be suggested by the polemics of the pro-life/pro-choice debate."

Noting recent rapid developments in medical technology including genetic engineering, IVF, cloning and the Human Genome Project, Day observed that we now had the ability to do things previously unthought of, such as detection of foetal abnormalities. But should we?

In relation to the moral status of the embryo, while many Christians speak as if the Bible settles the issue, Allan said: "there is no basis in Scripture or science to consider every embryo as a person, that is, with a moral claim on us." Allan agreed with Professor Gareth Jones (Otago University) that the embryo should be considered as a potential person.

Allan's stem cell paper was written when Australian federal legislation (2002), which allowed surplus (IVF) embryos to be used to produce embryonic stem cells, but not the creation of embryos for such a purpose, was being reconsidered. Many conservative Christians opposed the call from the scientific community to allow therapeutic cloning, because embryos, which they regarded as persons, would be destroyed. (The law was in fact liberalised in December 2006.)

Allan Day, in humility, recognised that "a variety of biblically sustainable positions may be held by well-motivated Christians." He also said, "one might recognise, but again not necessarily share, the unbending opposition of conservative ethics to abortion under all circumstances even if other lives are considered or if pre-implantation embryos are involved ... this is perhaps a warning about the difficulties in reaching consensus ... One must deal sensitively rather than polemically with these issues."

The Human Story

He saw clearly that if we put our views harshly, stridently, contemptuously or arrogantly, either within the church or in dialogue with the world, the effect will not only probably be counterproductive, but we will hardly be imitating and modelling the character of Christ, which he took to be at the heart of Christian ethics.

Conclusion

We will all miss Allan Day's forthrightness and his insightful grasp of the issues at stake at the faith–science interface. He ensured that the conversation was vigorous, rigorous and focussed on a search for the truth. We thank God for his contribution as an outworking of his Christian discipleship.

Towards the end of his life Allan struggled with his failing hearing, sight and arthritis. These prevented him from being as productive as he would like to have been. Now, for Allan, the wait is over, and we are sure he will have heard, "Well done you good and faithful servant."

> This chapter is a shortened version of presentations to the ISCAST Victoria meeting, 23 November 2013. The full versions of the presentations are available at www.iscast.org.

39. Stephen O'Leary: Using science as a tool in the Lord's service

By Stephen O'Leary

In the following testimony, the author describes his Christian journey and calling as a scientist, clinician and researcher (under the mentorship of the inventor of the bionic ear, Professor Graeme Clark), and how music has also nurtured his Anglican faith.

When I was very young, the stars and astronomy inspired a profound awe in the wonder of creation. I would spend hours staring through my telescope at the planets, and at galaxies or star clusters where the light had taken hundreds of thousands, and in some cases, up to millions of years, to reach earth. Man had just walked on the moon, and the extraordinary courage and optimism of this feat had a profound effect upon my thinking.

From the perspective of an eight-year-old boy, it seemed that anything could be achieved if there were a will. Little did I know, or care, of the politics of a cold war that arguably motivated this moment in history. But I still believe, and very deeply feel that the amazement of the world's people, who literally stood still when Neil Armstrong and Buzz Aldrin walked on the moon, transcended the most cynical interpretation of why humanity was there. But more than just amazement, the moon landings led me to believe that scientific endeavour was a profoundly pure enterprise. Something worth dedicating one's life to.

It was soon afterwards I joined the local Anglican church choir of St Martin's, Killara in Sydney. We were very lucky that the choir master was one of Sydney's best organists, the young (at the time) Peter Kneeshaw, who was winning eisteddfods, and who would later serve at Christ Church, St Lawrence and then St Mary's Cathedral for more than 20 years. What an experience it was to listen week in, week out to a truly gifted musician interpret the world's best organ repertoire. But more than that, I grew to love choral music, and to appreciate the splendour of the musical tradition of the Anglican Church.

I came to appreciate that God had spoken through different languages over the ages, but that the message of the gospel had remained the same. That music that had survived had done so because of its quality, beauty and enrichment of people's worship. These experiences sowed the seed for music playing a pivotal part in my faith journey.

The Human Story

Arguably the defining moment in my career came when, as a medical student, a colleague from the Christian Union at the University of Melbourne, Chris Walters, invited me to the Royal Victorian Eye and Ear Hospital to see some research that he was undertaking. This turned out to be with Professor Graeme Clark, who was at the time just beginning human trials with the bionic ear. Before the afternoon was out, I had been interviewed by Prof. Clark to undertake a similar research project the following year. During my research, the first commercial cochlear implants were undertaken, which I with others from the department watched via a remote video link.

It is hard to describe the intensity of the emotions that resonated through our university Department of Otolaryngology that year. The feelings of optimism, anticipation, excitement and, once the cochlear implant was successful, sheer joy, were palpable. Once again, I had experienced what purpose, scientific creativity and dogged determination could do. As a young and idealistic Christian, these experiences strengthened my personal resolve to dedicate my life to improving the health of others, not just through the practice of medicine, but also through medical research. Fortunately, I was also to learn that I had an aptitude for the latter.

Until this stage in my life, decisions had been relatively easy. The hardest had been deciding upon whether to study medicine at university, or follow my initial desire to read science. I had been brought up in a Christian family and had been fortunate to be led to a good youth fellowship at St Peter's in East Lindfield in Sydney. Upon moving to Melbourne with my family, I had embraced the Christian Union at the University of Melbourne and whilst an undergraduate had come to appreciate the breadth of Christian philosophy and expression. But the time was approaching when the decisions were to become much more difficult, and would likely have life-long ramifications. Discernment of God's will for my life would be even more necessary.

The first of some major decisions was whether to undertake a PhD several years after graduating from medical school. This alternative was seen by many as a gamble—to leave hospital practice might mean that opportunities to undertake a medical specialisation would be lost. But I felt led to take this course. God had presented life opportunities that had revealed the power of scientific pursuit, and shown that I had at least a little talent, and through an openness to hearing his calling through prayer the PhD was undertaken.

A full-time PhD is a great privilege. It is, arguably, the only time in life when one can be completely immersed in a problem, and not have to juggle

Stephen O'Leary: Using science as a tool in the Lord's service

this with other pressing issues, such as clinical work or administration. For me it was a joy, no doubt somewhat heightened by courtship and marriage to my wife, Catherine. But the PhD also allowed me the time to sing. I was by now taking singing lessons, and joined Trinity College Chapel Choir (Melbourne) under the direction of Bruce McCrae.

Chapel choirs specialise in providing the music for Evensong, which I have come to appreciate as one of the most contemplative of services. It reflects upon the day, and prepares for the evening. It takes the congregation away from the business of the work that has been done, calms the spirit and brings worshippers into a place where they can be receptive to God's spirit. And while music may not be the service, it is a potent instrument to bring the congregation into an attitude of contemplation.

Some of the most beautiful music in the church music repertoire has been written for Evensong, and it is my experience that the Scripture behind the text is brought to life through it. The Scripture in music lingers longer in the mind than text that has been read, and more so, brings an emotional connection that can lead to a richer spiritual understanding.

During this period of my life I was also exposed to a wide repertoire of early liturgical music through the Tudor Choristers and Ensemble Gombert under John O'Donnell. One has to work harder for the spiritual rewards from this music because much of it is written in Latin, but when placed in a liturgical context the task is made much easier. And if one is open it is possible to glimpse how worship may have moved our sisters and brothers in Christ from previous generations.

I am struck by the sensitivity and adoration in music written about the life of Jesus' mother Mary. Mary had not held prominence in my church experience, but the music written in her honour gives me a glimmer of understanding of the esteem in which she was held by a believer in the fifteenth century (or others from different Christian traditions today).

When the opportunity arose to undertake post-doctoral research at Oxford my wife Catherine and I jumped at the chance. We were newly-wed and without children, so we relished the prospect of living in the UK, in close proximity to London. Soon after arriving, I had joined the occasional (relieving) choir of Christ Church Cathedral and when possible attended Evensong at New College. We were travelling to London several times a week to attend concerts at the Barbican, South Bank or the Proms.

But a big decision was ahead—whether to remain in research or specialise in ENT (Ear Nose and Throat) surgery, so that I could treat ear disease. This too was not a straightforward decision. Some at Oxford warned that once a young scientist left the laboratory they would struggle to get future jobs in the field. It seemed as though I were on an ocean liner, and thinking of surgical training was like turning the ship around. And then there was the harsh reality that selection into surgical training required passing an examination with a 10 percent pass rate, and then being offered a job. There was uncertainty whichever way one looked, and prayers at this time asked the Lord for guidance but also for courage.

Surgical training and children followed, and finally a return to Melbourne to take up an academic position in ENT at the University of Melbourne and Royal Victorian Eye and Ear Hospital. We had lived in three UK cities, Sydney and the Netherlands. Our eldest child had had 14 homes by the time he was six years of age, and had had a year in a Dutch-speaking school. All families know the busyness of those years with young children, and for ours it was also that of surgical training.

It is interesting to reflect upon what that meant for our Christian walk. We had experienced the gambit of welcome in the churches that we had attended, sometimes for as little as six months. I think that Catherine and I came to appreciate more deeply that true Christian welcome is more concerned with the needs of the others than ourselves; those who extended the hand of friendship to us, even though we may have only attended their church for a short period, were great exemplars.

The constant moves meant also that Catherine and I needed more than ever to support each other in our Christian lives, and not just for us as a couple but also for our children. Personally, I took great comfort in the Anglican community, which gave us a common connection during our travels. My musical journey also took a very different path. We were now attending churches where contemporary worship was the norm. Now, the music making was a much more corporate affair, and engagement with the music was quite different. The congregational experience of the hymn accompanied by organ was now the form of popular music, created by members of the congregation and shared by all. But I have come to see that this form, particularly with its repetition in chorus and bridge has its place in the teaching of Scripture and helping people to contemplate upon God's word.

Personally, what I love most is to extemporise harmonies and to be sensitive to the Spirit moving in the congregation. (Followed not far behind by the excitement of youth who are participating in their first band.) The music is in some ways akin to the great traditions that I have loved all of my life, in that worshippers finally enter a personal communion with the Lord as they reflect upon the text. Perhaps the glory of a renaissance motet may have evoked similar joy for our forebears as does a well-written contemporary song, delivered by a church band open to God's leading?

The last major point of decision in my career (to date) was whether to apply for appointment to the Chair of Otolaryngology at the University of Melbourne. To take on this responsibility was never going to be an easy task; the constant need to generate funds is a matter of survival not just for the department, but for the livelihood of those within it. The complexity of life would increase considerably with administration and university politics to contend with. But this role is primarily about driving the research strategy for the discipline, and my desire to improve the understanding and treatment for ear-related disease had not abated. Was it God's will that I should apply for the job? Prayer for guidance, for an openness to discern whether this was God's will was particularly important at this time.

I came to understand that the Lord had presented me with life experiences that had driven a passion for enquiry, a desire to make a difference, and through my subsequent career it appeared that I had been blessed with the capacity. The chair was a vehicle where these gifts could flourish. Trusting the situation to the Lord, I applied and was appointed in 2008. For the clinician-scientist, the drive for research always starts with the patient, with the problems that we cannot solve; it is the individual clinical encounter that firms a resolve to take a particular problem into the laboratory. But discernment is required when setting a research agenda through to designing the experiments; there are always more choices than resources, and the art of science is to take the right turn. I continue to seek God's guidance in this, and for the courage and resolve to arrive at bringing new treatments to the clinic.

Having been given the privilege of working with science, how do I view it within the context of faith? I view science as a tool for testing and furthering our understanding of the world. Science creates a conceptual framework where reproducible observations can be understood, and the comprehensiveness of that framework tested. It does not concern itself with matters of faith, and it is my view that those who try either to prove or to disprove the existence of God with science are not appreciating the limits of the discipline. For these reasons I have no difficulty reconciling the use of science to improve the plight of those with human disease—in other words to use science as a tool, much like my surgical instrumentation—to learn how to treat patients better. In many ways I equate contemporary clinical practice with the scriptural (Spirit-given) gift of healing, while the use of science to improve health with the gift of creativity, all driven by service and compassion. These activities are deeply rooted in faith, and in our hope in Jesus Christ, our Lord.

Science and Christian Faith:
The State of the Marriage

40. The Christian conflict with science is dead

By Chris Mulherin

People of faith must resist the lure of the ghetto and engage with science if the world is to progress and if Christianity is to be a credible part of it.

In an increasingly global and secular scientific culture, the cutting edge of Christian engagement is the conversation with science. In fact, the progress or decline of Christian faith in the twenty-first century depends in large part on its dialogue with science.

For Christians, the current cultural skirmishes might seem to be about the best expressions of human sexuality or religious education in schools, but there is an underlying issue that those ones depend on and which is far more important; it is the prior question of whether the Christian faith can even be taken seriously in a scientific age.

In every generation, cultural and intellectual realignment redefines the "plausibility structure" that determines the limits of what is credible, of what is believable, of what is even possibly true. And the task of Christian apologetics—the defence of the faith—is to enter the cultural fray and argue the case that the Christian faith is a credible worldview. No amount of discussion about marriage, for example, is relevant if Christianity's claim to truth is written off as hocus-pocus. And if science is the norm of truth, then the credibility of the faith depends on the way people view its relationship with science; if people are convinced that there is a fundamental conflict between science and religious belief, there are no prizes for guessing which side most will vote for. So, while Christians are confident that "the gates of hell will not prevail" against God's church, that is no guarantee of a continuing cultural majority. Nor is it a theological excuse for retreat from the marketplace of ideas.

Yes, the faith will endure. But "love God with all your mind" and being prepared to "give an account for the hope that is in you" amount to a biblical call to Christians to engage vigorously with the powerful voices that would sideline Christianity without taking it seriously. In G. K. Chesterton's famous quip, Christianity "has not been tried and found wanting; it has been found difficult and left untried." Although "the conflict thesis" is no longer taken seriously at the level of thoughtful discussion, many public perceptions are driven by both political correctness (so faith is left untried) and what the philosophers call *confirmation bias*, which means that when faith is found difficult, it is rejected because it challenges one's deeply held views rather than confirming them. And for those who find faith difficult and who would rather leave it untried, it is overwhelmingly to science and the conflict thesis that they turn for solace and confident unbelief. In sociological and intellectual terms, the science–faith conversation is the cutting edge of Christianity surviving in the Western world; it's the front of advance or retreat of credible Christianity.

The gifts of science and of faith

Both science and Christianity are gifts of grace, either of which we disrespect to our peril.

The gifts of science are numerous, breathtaking and worthy of deep gratitude. As a means of discovering truth about the natural world, science is outstanding, offering extraordinary insight into the mechanisms of the universe and of life itself. Scientific knowledge offers a power that has led to rapidly increasing health and wealth for all, including the poorest of the global population. And despite continued inequity, as well as the abuse of the power of science to commit appalling atrocities, such blights cannot be blamed on the scientific enterprise. Why? Because no amount of science can provide answers to questions of meaning or morality. Science cannot tell us when its products are well spent and when not. It cannot tell us if the means of ending life painlessly should be used; it cannot tell us whether the next generation of military weaponry is for good or ill; it cannot tell us whether we ought to spend billions on space exploration or sustainable agriculture. These are questions outside the ambit of science.

And the gifts of Christianity, too, at a purely secular level, are also manifest. Human rights entrenched worldwide, convictions about charity, compassion, justice, the social welfare net, equality—all have roots and motivations deep within the Christian faith. But the Christian worldview—so foundational to a Western culture of equality and corresponding rights—is being dismantled piece by piece. While vestiges remain, such as the equal dignity of all human beings or "do unto others as you would have them do unto you," they are now adrift from their roots, which lie in the conviction that humans are made in the image of God.

The Christian conflict with science is dead

Culture and credibility

With globalisation and the spread of techno-scientific thinking and practices—most obviously exemplified in the internet—a secular scientific worldview is advancing to all corners of the earth. This view, most aggressively championed by the so-called New Atheists, challenges all non-scientific thinking in its advance.

Today the right to be heard depends partly on getting along with mainstream science. And, in a sense, that is as it should be. But "mainstream science" is not the same as *scientism*, the ideology that says that science has (or will have) all the answers. Scientism, which goes well beyond healthy science, is becoming the cultural default position: "When it comes to facts, science is the only game in town," says Daniel Dennett, philosopher and one of the media personalities of popular atheism. So the comprehensive Christian worldview, which has for centuries included science as an essential element, is increasingly dominated by a view that science, seen as the epitome of sure knowledge, offers the only access to truth.

In the face of this changing balance of cultural forces and views about what is credible and what should be relegated to incredibility, there are two options open to Christians.

The first option is to beat the retreat to the Christian ghetto. This path would accept that science and all religion are worlds apart and that—borrowing from Os Guinness—Christianity might be privately engaging but it is publicly (and scientifically) irrelevant. This is a way backward; it involves denying that Christian faith is true in any serious sense. It involves accepting the New Atheist line that faith in Jesus Christ is akin to believing in the tooth fairy or Father Christmas. It also involves ignoring the biblical record, including the words of Paul that if Christ was not raised bodily from the dead, then our faith is in vain.

But there is another option: a way that has been the orthodox manner of engagement since the beginning of the Christian era. Following the example of the Galilean teacher, Paul the apostle debated with the public world of his time on the Areopagus in Athens—also known as Mars' Hill. And for 2000 years since, thoughtful Christians have proclaimed that the God of the Bible is revealed both within that book and also through achievements of the arts and sciences. This second option is to follow the path trodden by the great Christian scientists and thinkers of history and to thoroughly affirm the two books of God—the book of his word and the book of his world. In every generation, it needs to be proclaimed again from pulpits and peer-reviewed articles: there is no conflict between science and faithful Christian belief!

The conflict thesis is bunkum

The past crowd of witnesses who saw no conflict includes hundreds of the great names of Western history. To name only a few who are prominent in the history of science: Roger and Francis Bacon (linked across three centuries by their names and by laying foundations for the scientific method), William of Ockham (remember Ockham's Razor?), Jean Buridan (presaging inertia), Copernicus, Kepler, Galileo, Descartes, Pascal (remember his triangle?), Boyle (of gas law fame), Linnaeus (natural taxonomy), Bernoulli (his law keeps planes in the air), Lavoisier (we owe chemistry to him), Faraday (invented the electric motor), Maxwell (electromagnetic fields), William Bragg, Max Planck, Werner Heisenberg (creator of quantum mechanics)— the last three also being Nobel laureates.

And, in case you are prone to the prejudice that devalues past thinkers as if they were semi-incompetent and ignorant in the light of present knowledge, there is no question that numerous outstanding living scientists and respected thought leaders—many of them mentioned elsewhere in this book—are also Christians: John Houghton, lead editor of an early report from the United Nations Intergovernmental Panel on Climate Change; Francis Collins, who led the human genome project and is now head of the US Government National Institutes of Health. And in Australia, people such as Graeme Clark of bionic ear fame or Glenn Stevens, past head of the Reserve Bank, are household names, and people for whom Christian faith and serious engagement with "secular" thinking poses no conflict.

Of course the point is not that any list of Christians who are also people of public stature proves that Christianity is true; what it proves is that it is possible for publicly respected leaders who are rigorous thinkers to be committed to the truth of Jesus Christ.

Philosophers of religion, too, are aware of this and while they may personally be atheists, they often recognise that there is no necessary conflict between science and faith. Atheist philosopher Jim Stone says believers are often excellent philosophers and respected by their atheist colleagues. But he is frustrated with fundamentalist atheism, which has no conception of its own blinkered approach. He says: "The people I don't like are the New Atheists, because they don't seem to realise that the [Christian philosophers] with whom I must contend even exist." Other prominent thinkers too are frustrated by hysterical atheism. The un-reasonable and vitriolic attitude of books such as Richard Dawkins' *The God Delusion* prompted atheist philosopher of science Michael Ruse to say that the book makes him embarrassed to be an atheist.

Into the fray

The time for simplistic belief and unbelief is over. Fundamentalists, religious and atheist, must give up their ground to views that hold science in its rightful place as servant of a broader worldview—Christianity in its fullness—which offers the soil out of which grew both modern science and a global framework of equality and human rights.

The need is for Christian thinkers, and especially those who are involved in science and technology, to take up the gauntlet laid down by secularists and to speak up and to speak loudly about their own experience of integrating their faith with the best that science has to offer. Christian scientists must come out of the shadows. Their science is important but the future of a culture deeply rooted in human dignity and meaningful existence depends also on knowing there is more to truth than what science can offer.

The need is for pastors to convince their flocks so that no Christian lives with that uneasy secret suspicion that faith is actually the antithesis of science and serious thinking.

The need is for theological educators to ensure that students comprehend that the study of God's word and the study of his works (the creation) are not incompatible exercises; the basics of apologetics and some understanding of science should be par for the course in training for Christian ministry in a scientific age.

In short, the need is to proclaim in every pulpit and public space, in academia and the Twittersphere, in every Christian classroom and lecture theatre, that the conflict thesis is dead—in fact, never was a credible view—and that the survival of human society as we know it depends on a healthy relationship between science and Christianity.

41. Is life without purpose, and religion irrelevant?

By Mick Pope

It is a widely held belief that religion has become irrelevant under the relentless advance of science. In particular, religion is a crutch for the weak, offering meaning and purpose where none may be found. In his book *The Hitchhiker's Guide to the Galaxy*, Douglas Adams offers an illustration of this with a scene in which a nuclear missile is accidentally and improbably turned into a sperm whale: "Ah ... ! What's happening?" it thought. "Er, excuse me, who am I? Hello? Why am I here? What's my purpose in life? What do I mean by who am I?"

The answers to the whale's comical questions are: you are an accident, a cosmic fluke and your pointless life will soon end by crashing onto the planet below. Richard Dawkins in *The God Delusion* claims Adams as his disciple in the meaninglessness of life, having been influenced by Dawkins' earlier book *The Selfish Gene*. The whale is a cipher for us all in the meaninglessness of existence.

Challenges to meaning

There are four broad scientific challenges to the ideas of purpose and meaning. The first comes from evolution. Evolution by natural selection sifts the results of random mutations in genes, passing on those that make organisms fit their environment. Stephen Gould in his book *Wonderful Life* claims that natural selection does not follow a "ladder of predictable progress" inevitably leading to humanity but instead is a "copious branching bush, continually pruned by the grim reaper of extinction." We are not special nor made in God's image but merely "a naked, upright ape" that would not appear if the "tape of life" were re-run. We are not even the only intelligent species on the planet. Crows exhibit forward planning; gorillas experience grief and cuttlefish have extremely complex ways of communicating.

A second challenge comes from cosmology. Why is there a universe at all? In his book *The Goldilocks Enigma*, Paul Davies explores ideas offered in place of God. Our universe is said to be part of a larger multiverse, an "eternal bubble-bath universe" from which baby universes spontaneously come into being due to the laws of quantum mechanics. One of these baby universes is bound to be suitable for the appearance of life. In such an infinite multiverse it is suggested that civilisations could arise that could create entire universes inside computers so that we live in "the Matrix," or where God himself has evolved along the lines of the *Q Continuum* from *Star Trek: The Next Generation*, omnipotent, omniscient but nonetheless creaturely and flawed. There may even be infinite copies of you and me!

A third challenge is the problem of evil and suffering in nature. Evolution appears to be a thoroughly wasteful process. Over 98 percent of all species that have ever existed have gone extinct. Sometimes extinction happens slowly and on a small scale; at other times, the entire life on earth has been threatened. At the end of the Permian period, as many as 99.5 percent of all organisms died. This says nothing of the day-to-day horrors of predation, parasitism, infection, floods, lightning and earthquakes, all of which appear to be part of the backdrop of existence.

The final challenge comes from the fact that everything has an end. We all die, which for some denies any purpose to human life (Ecclesiastes talks about this at length). The sun's fate is to expand and grow hotter, boiling away the earth's oceans in about four billion years' time. Perhaps humanity will be able to escape to other solar systems, maybe even other galaxies. In the end, however, all stars die, becoming black holes, neutron stars or white dwarfs. Even these decay over time until the ever-expanding universe becomes a cold, inky nothingness some 10^{100} years into the future in the so-called heat death.

These are not small issues and responses invariably risk sounding trite. However, a thoughtful Christian can respond with answers that go beyond "because the Bible says so." Further, given the freely speculative nature of some of the above discussion on the part of scientists, there is plenty of room to move in addressing the issues.

Room for the creator

For many, creation and evolution present a choice. Either God did it or it happened all by itself. This is a false dilemma. God has given creation the capability to realise itself, to bear fruit in the varied and wonderful ways which we observe. Recent evidence for this comes from Simon Conway Morris's book *Life's Solutions: Inevitable Humans in a Lonely Universe*. Morris sees evidence for the phenomenon known as convergence, the process where similar physical features arise independently for very different creatures living under similar evolutionary pressures. For example, "tigers" with sharp incisors for ripping into prey have evolved among both cats and marsupials. Different sorts of eyes have appeared independently.

Most significantly, intelligence has appeared among a number of different species (crows, whales, cuttlefish and primates) leading Conway Morris to suggest that the universe was destined to produce mind.

Furthermore, it is remarkable how often biologists have had to "remind" themselves that what they see is not designed and to avoid such language. Design is not an illusion of the mind, but the perception of the hand of God. Creation is both an emergent and an elective process. It is emergent in that genuinely new things arise "on their own" that cannot be simply reduced to that from which they are made. For example, life is more than the chemicals that make it, consciousness more than the neurons firing in the brain. However, creation is elective in that while nature doesn't need God to "interfere" with it because he has made it fruitful to realise its own purposes, he can choose to interact with it. Human existence is made possible by divinely inspired creation, and made definite by divine action. This election occurs because God chose humanity to bear his image (see Psalm 8).

Arguments about eternally existing multiverses and such are simply substitutes for God. Why is there an infinite multiverse? Why do any of the baby universes contain life? In the end, either positing God or an infinite multiverse is a non-scientific choice based on a metaphysical position. However, which assumption makes the most sense of existence? In this case, beauty, design and ultimately the empty tomb point in one direction.

Beyond the end

The Bible stresses a future physical existence, a resurrection from the dead for a redeemed humanity and a new heaven and *earth*. The problem for this comes from the second law of thermodynamics (the law of entropy) which essentially says that everything is slowly running down. Yet, the second law is not a fundamental law of nature like gravity but a statistical description. Hypothetically speaking, the pieces of broken coffee cups can rise from the floor to a table and come together again, if you wait long enough (although you might be waiting many times longer than the age of the universe). Could God transform this creation so that entropy runs backwards? Does the universe show signs now that it is changeable in a way compatible with a "new heavens and earth"? Early in the universe's history, the four fundamental laws emerged out of one unified law. Could such a change happen again in the future? Physics provides us with enough openness to freely speculate and not to close down the debate too early on the side of futility and purposelessness to existence.

Is life without purpose, and religion irrelevant?

Christians look forward to the resurrection, when every tear will be wiped away, yet what is often overlooked is that Paul writes about the entire creation being redeemed (Romans 8). The details are unclear, but suffering is to be dealt with. Even agnostic Michael Ruse in his book *Can a Darwinian be a Christian?* acknowledges that the crucifixion, the idea of a suffering God, offers hope to a suffering creation.

The prime characteristic revealed about God in the Bible is not philosophical qualities such as transcendence, omnipotence and omniscience, but love. God creates to include something outside of his triune self into the divine relationship. In love, God allows the other to be and to develop according to its potential. This is achieved in a universe that can evolve from simplicity to complexity, from non-life to life and from non-consciousness to consciousness and hence onto image bearing. This process appears unavoidably to involve pain and suffering without God directly causing that suffering. Furthermore, this letting be suggests self-imposed "limits" on God's other characteristics such as omnipotence and omniscience. The divine act of creation is self-emptying in exactly that sense that God limits God's own self by allowing the other to be. The biblical story holds this in tension between a God who is in control of the destiny and purpose of his creation.

Questions about ultimate purpose and meaning take us to the heart of what it means to be human. Science has not, and ultimately cannot close down this discussion. There is still room for speculation, for philosophy and theology.

42. A "more subtle science" reveals divine reality

By Colin Goodwin

Distinguished Australian scientist Charles Birch, who died in 2009 at the age of 91, believed there is no conflict between genuine science and genuine religion. Colin Goodwin explores Birch's thesis that science and religion belong together in a single, unified worldview.

John Habgood, former Archbishop of York (and distinguished Cambridge researcher in physiology), once splendidly referred to Charles Birch as "an ebullient polymath," and a person who "writes as he lives: with passion, elegance, and humour." This ebullient polymath, born in Melbourne in 1918, was a world-ranking biologist who, he tells us, was "brought up as a low-church Anglican," and subscribed to "a total package of fundamentalism." He belonged to Christian evangelical groups both at school and at Melbourne University, from which he graduated in agricultural science in 1939. Graduate research at the University of Adelaide was accompanied by a shattering of his fundamentalist outlook and his discovery, via the Student Christian Movement, of "an alternative interpretation of Christianity to fundamentalism."

Following two years of research in the United States and England, Birch commenced work at the University of Sydney in 1948, becoming Professor of Zoology in 1960, and Challis Professor of Biological Sciences from 1963 until retirement twenty years later. A member of various international scientific societies, he also served for twenty years on science-related committees of the World Council of Churches. Charles Birch, author of six books, co-author of another six, and author of numerous scholarly articles, was in 1990 awarded the prestigious international Templeton Prize in Science and Religion.

The question of how science and religion stand in relation to one another is a central motif of what Charles Birch has to say to the contemporary world. He fully endorsed the comment of the great Anglo-American philosopher Alfred North Whitehead that "When we consider what religion is for mankind, and what science is, it is no exaggeration to say that the future course of history depends upon the relations between them." Birch identifies three major ways in which science and religion are related. The

A "more subtle science" reveals divine reality

first is one of conflict, of contradiction: demonstrably valid knowledge of the world comes only from the recognised physical sciences—physics, chemistry, biology, geology, etc. No such knowledge comes from religion, the claims of which are to be rejected as not being backed by adequate evidence. The second possible relationship between science and religion is one of contrast, of "running in parallel": science explores questions of "what" and "how" regarding the physical universe, whilst religion asks parallel questions about the "why" of human and cosmic existence, i.e., questions about meaning and value.

Charles Birch is unhappy with both of these accounts of the science-religion interface. Both take for granted mechanistic and "substantialist" views of the entities that compose the universe. That is to say, both accept a universe that is to be fully interpreted in terms of bits of matter (sub-atomic particles, atoms, molecules etc.) being moved in space, and physically interacting on the basis only of deterministic natural forces. These bits of matter or stuff "remain unchanged no matter what particular whole they constitute, be it a star or a brain."

The third way in which science and religion can be related is what Birch calls the "integration thesis." He means by this that science and religion belong together in a single, unified world-view, reflecting the fact that "The mind that is involved in religious concerns is not a different mind from the mind involved in scientific issues." Birch offers an account of this integration that is quite breathtaking in its conceptual range, richness and rigour.

The first, and absolutely indispensable thing that an integrated science and religion have to do is to recognise the world as it really is. The world with which both science and religion are concerned is not just a world of static substances, of bits of matter or stuff that interact only externally through physical forces. It is a world endowed with what Birch calls "subjectivity" or "mentality" that belongs to all individual entities without exception. He means by this the intrinsic capacity of each individual entity to take account of its situation, to respond to it in its own way experientially from within itself—from the "inside."

The best example of this capacity—one immediately known to us—is, of course, our own self. We experience our own subjectivity from within, i.e., our own intrinsic capacity as human beings to take account of our situation, to respond experientially in self-determining fashion to what we encounter. Birch argues strongly in all his writings that this capacity extends analogously or proportionately in an unbroken line from humans down to electrons and quarks. "The ultimate entities of the universe," he tells us, "are subjects with their own degree of spontaneity, self-determination, freedom, and sensateness or experience. This is the doctrine of pan-experientialism." This is to perceive the world "in depth," in a way that goes beyond "the external and statistical aspects of events" that are the concern of "science as now practised." It is to perceive the world in a way that bridges the gap between the data of science (e.g., an electrical current in a nerve cell) and the data of our lived subjective experience (e.g., a feeling of pain), and is a way of perceiving that extends to the subjective data of all other created experiencing entities as well. Nor is Birch slow to call for "a recognition by science of a wider field of exploration that includes the subjective ... the time has come for science to march to another drummer."

At various points in his writings Birch sets out lengthy analysis and argument to show that the existence of subjectivity in all the individual entities of creation requires the existence of cosmic mind as the very ground of the universe. Cosmic mind or God as identified by Birch is not an "omnipotent, supernatural, legalistic ruler of the universe"—an "interventionist" God, as he often puts it. As understood by Charles Birch, God is the pervasive ground of all the possibilities that would enable any universe at all to exist, and the one who persuades or "lures" possibilities/potentialities into actuality. (In a vivid phrase of Whitehead, God is "the unlimited conceptual realisation of the absolute wealth of potentiality.") Drawing on Whitehead's terminology, Birch sees this ontological pervasiveness as God's "primordial nature," and the mode of God's presence in the world. God's "consequent nature" is the presence of the world in God, together with God's conserving of all created values—the doctrine of panentheism (of all-things-within-God). For Birch, "What is of value in all existence is somehow saved as memory forever in the life of God." This has the effect of "establish[ing] the real importance of all we are and feel and all that the rest of creation feels." This position implies, in turn, that "Our existence and that of the rest of creation enrich the divine life, and that is the ultimate meaning of existence. The world is saved."

In discussing the interface or dialogue between science and religion, Charles Birch thinks, and writes, as a practitioner of science and as a Christian ("though one who may not measure up to the judgment of some who regard themselves as the trustees of orthodoxy," as he once ruefully noted). He looks explicitly to the opening chapter of the Pauline Letter to the Colossians to discern "the theme of Christ and 'all things,' the theme

A "more subtle science" reveals divine reality

which sums up a reasoned faith that can make sense of science as of life." For Birch, Christ is the highest exemplar of the experiencing subjectivity belonging to every individual created reality. In him the "pleroma" or fullness of life is said to dwell (Col 1:19). In a striking piece of exegesis (of Colossians 1:15: "[Christ] is the image of the invisible God"), Birch affirms that "in all creation, there is nothing bigger, nothing higher, nothing more fully manifesting the nature of God than Jesus Christ. If you do not see that, [the Pauline writer] is saying, you miss the whole point of the universe. This is where you start in your understanding of all creation." Charles Birch most certainly does not "miss the whole point of the universe": it is always finally to reveal the true nature of divine reality. And all of his reflections on the science–religion interface are pointed ultimately towards a reasoned disclosure of this divine nature.

"We are participants in the adventure of an unfinished universe," Charles Birch reminded us and

> there are no real battles between genuine science and genuine religion. There will always be battles when one or both are bogus: when dogma replaces interpretation based on experience, or when certain experiences are excluded *a priori* as, for example, when the mechanist excludes anything that cannot be weighed or measured.

The work of Charles Birch is to be welcomed as a brilliant response from an Australian to Whitehead's famous call for "a deeper religion and a more subtle science."

43. A brilliant exploration of the science–religion debate

By Colin Goodwin

A review of *Science & Religion: A New Introduction (Second Edition)*, by Alister E. McGrath (Wiley-Blackwell, 2010).

Let me say at once that the more I read of Alister McGrath's exceptional output of theological writing the more I am tempted to say that he is to twenty-first-century theology pretty much what Wolfgang Amadeus Mozart was to eighteenth-century music. In terms of broadness of scope, sustained brilliance, variety of expression, and challenging nuances, these two giants of high culture reveal in their respective domains a striking degree of similarity.

The second edition of Alister McGrath's *Science & Religion: A New Introduction* underscores this point. Any of an array of laudatory adjectives might be employed to evaluate the book. I shall employ *wondrous* in its standard sense of "causing amazement and admiration" to do so, just as one might use the same adjective to praise, say, Mozart's flute and harp concerto in C, K. 299. That McGrath's *Science & Religion: A New Introduction* is capable of "causing amazement and admiration" is clear for at least the following reasons.

Taken as a whole, the book is a superbly structured compendium of practically everything important in the discussion between science and religion, with three great debates—Copernicus, Galileo, and the Solar System; Newton and the universe as a vast mechanism; Darwin and human biological origins—providing a lively prelude to later history of the relationships between science and religion. A recurring topic here is, of course, biblical interpretation. As McGrath notes, "Conflict between science and religion often arises when scientific advance is seen to conflict with the prevailing modes of biblical interpretation"—something pointedly evident in the first and third of the debates just mentioned. Theologian Ian Barbour's typology of "ways of relating science and religion,"—conflict, independence, dialogue, and integration—is extensively (though not uncritically) drawn upon by McGrath in his unwavering attempt to understand these fields.

In Part 2 of the book McGrath covers what he calls "twelve general themes of interest to the dialogue between science and religion." This fascinating section explores issues such as the notion of "explanation" in

A brilliant exploration of the science–religion debate

science and religion, verification and falsification in these areas, proving a scientific theory and proving God's existence, the doctrine of creation and the natural sciences, and how non-Christian belief systems (McGrath considers Judaism, Islam, Hinduism, and Buddhism) approach science–religion interaction. Behind much of what is said regarding these themes hovers McGrath's question: "What are we to make of the widespread belief that science *proves* its beliefs, whereas religion merely *asserts* its beliefs, often in very dogmatic ways?"

The third part of *Science & Religion: A New Introduction* deals with a number of matters relating to specific scientific disciplines in their bearing on religious thought. Cosmology discusses the universe's origin and the whole physical order's remarkable degree of "fine-tuning," with obvious resonances for religion. The discussion in quantum physics of "complementarity" (wave-particle duality) to appropriately model sub-atomic phenomena recalls at once the complementarity in Christian doctrine of "two natures"—one divine, one human—to model appropriately the phenomenon of Jesus of Nazareth. Evolutionary biology and religious thought continue stimulating (sometimes strident) exchanges around the notions of purpose and "design" in biological systems; whilst psychology and, more recently, cognitive science, engage with religious faiths about the origins of religious belief, the nature of religious experience, and human predispositions to believe in divine agencies.

McGrath concludes his book with a cluster of insightful chapters entitled "Case Studies in Science and Religion." These studies offer brief sketches of the careers, and salient arguments, of ten major recent or contemporary contributors to the science–religion interface, and run from Teilhard de Chardin to, for example, John Polkinghorne and Philip Clayton.

Like the music of Mozart, the theological writing of Alister McGrath is inexhaustibly wondrous. Like Mozart's flute and harp concerto in C, K 299, McGrath's *Science & Religion: A New Introduction* is altogether compelling in its capacity to cause "amazement and admiration."

44. Christianity defended with logic and vigour

By Colin Goodwin

A review of *Atheist Delusions: The Christian Revolution and Its Fashionable Enemies*, by David Bentley Hart (Yale University Press, 2010)

The Archbishop of Canterbury put his finger on the truth at the awarding of the 2011 Michael Ramsey Prize for theological writing to author David Bentley Hart, for *Atheist Delusions*, when he said the book "shows how the most treasured principles and values of compassionate humanism are rooted in the detail of Christian doctrine."

What Rowan Williams had in mind were the extended, entirely convincing, lines of argument pursued by the author of *Atheist Delusions* that took this impressive American theologian from doctrine about Christ elaborated at the great fourth and fifth-century councils of the church—doctrine finally formulated at Chalcedon (451 AD) that two natures, one divine one human, exist complete and undiminished in the one divine person of the incarnate Word—to quite staggering conclusions about human beings. These conclusions, thoroughly grounded in this doctrine, centre on the unconditional preciousness of each human person, on

> the human as such endowed with infinite dignity in all its individual "moments," full of powers and mysteries to be fathomed and esteemed ... made in the divine image and destined to partake of the divine nature ... resplendent with divine glory, ominous with an absolute demand upon our consciences.

In wider perspective, the Christology of the councils as part of "the rise of Christianity" produced

> consequences so immense that it can almost be said to have begun the world anew: to have "invented" the human, to have bequeathed us our most basic concept of nature, to have determined our vision of the cosmos and our place in it, and to have shaped all of us (to one degree or another) in the deepest reaches of consciousness.

Yet "the more vital and essential victory of Christianity lay in the strange, impractical, altogether unworldly tenderness of the moral intuitions it succeeded in sowing in human consciences."

Christianity defended with logic and vigour

These conclusions flowing from the reality of the incarnation pervade the insights that David Bentley Hart offers in *Atheist Delusions*. We have what he calls the "Christian interruption" or "revolution" involving "a truly massive and epochal revision of humanity's prevailing vision of reality." Such a revision will not, of course, "magically transform whole societies in an instant ... or entirely extirpate cruelty and violence from human nature." Nor does Hart excuse the appalling failures of so many Christians over the ages to live lives of genuine love and peace.

Hart does, however, come repeatedly, with crisp logic and exceptional vigour, to the defence of "historical Christianity" against the more egregious nonsense peddled by such well-publicised opponents of Christianity as Richard Dawkins, Christopher Hitchens, Daniel Dennett, and Sam Harris. Pointedly, he laments that "we have lost the capacity to produce profound unbelief," that "the best we can now hope for are arguments produced at only the most vulgar of intellectual levels, couched in an infantile and carpingly pompous tone, and lacking all but the meagrest traces of historical erudition or syllogistic rigour." The days of "genuinely imaginative and civilised critics"—Hart speaks of Celsus, Porphyry, Hume, Voltaire, Diderot, Gibbon, and Nietzsche—are now behind us.

Late in his book Hart asks whether a truly post-Christian culture will also become post-human, when "the human" is understood as an invention of Christianity. Yes, says Hart, noting that a culture devoted to "acquisition, celebrity, distraction, and therapy," one that is not unreservedly against nerve toxins, nuclear weapons, genetic engineering, the killing of "flawed" human beings, and that celebrates its own "bizarre amalgamation of the banal and the murderous," is well positioned to alter its "view of human nature, or even the experience of being human."

Read—then reread—these brilliant, often troubling, summary-defying 253 pages: you will not be able to see human beings, nor the cosmos they inhabit, in quite the same way again. And all because of doctrine about the Word that has been "made flesh," and has come to "dwell amongst us."

45. Christianity, science and rumours of divorce

By Chris Mulherin

The rumours of irreconcilable differences between faith and science are based on misunderstanding.

Like all lasting marriages, Christianity and science have had to work at their relationship. But despite rumours of conflict, science and faith can look forward to an enduring marriage as they work together pursuing truth.

The "conflict thesis," which is thoroughly debunked by both historians and philosophers of science, has relatively recently been revived by an alignment of special interests. This includes the media's hunger for conflict as well as other provocateurs who have little respect for either serious history or philosophy; I'm thinking of the new breed of would-be public intellectuals such as Professor Richard Dawkins, the high priest of the New Atheism.

It is true that history records conflicts between representatives of science and religion but these historical examples of disagreement do not amount to philosophical or theological incompatibility. As in a marriage, differences of opinion do not constitute irreconcilable conflict.

The only conflict thesis worth taking seriously is one that suggests that there is a necessary and fundamental contradiction between science and Christianity. In this chapter and the next we will see that rumours of irreconcilable differences are based on misunderstanding, principally about the nature of science.

But before turning to science, the first idea worth remembering is this: Christianity is a worldview (and science is not).

Christianity is a worldview: It's about meanings, not mechanisms

One of the dangers of referring to the "science–religion relationship" is that this description appears to set up a symmetry between two comparable entities: science on the one hand, and faith on the other. But science and Christian faith are not directly comparable because Christianity is a worldview while science is not and never can be.

A worldview is a set of ideas and beliefs that offers a coherent framework to interpret the universe. It's a sketch of the "big picture." It answers such questions as: How should we live? How did it all begin? What happens after death? Does God exist?

As a worldview, orthodox Christian belief includes a "supernatural" creator God who made the universe and everything in it, the death and resurrection of Jesus Christ and the linear nature of history from creation through to final consummation. Christianity also includes an understanding of the purposes of a humanity made for relationship with God.

One implication of this description of the Christian worldview is that it is answering questions about meaning and not mechanics; questions about the purposes and not the particles of the universe. So Christianity is not directly comparable to science because science is not a worldview and—my second point—Christianity is not science.

Christianity is not science

The natural sciences—think of physics or biology or astronomy—search for the mechanisms and laws of the universe in the hope of answering the "how" questions. They look for the physical causes and constituents of what goes on in our world. Christianity is different.

On the one hand, as a worldview, Christianity is much more encompassing than science because it answers the big questions such as those above. But, on the other hand, Christianity has little interest in the "how" questions. For example, while the Bible tells us the meaning of the church and offers some general principles, there is no description of the mechanics of setting up the perfect church. Or, in the area of moral guidance, the Bible offers a general foundation but it does not tell us how to run a country or order our finances.

So Christianity is not science and it is a mistake to think that the Bible is a political treatise or a scientific textbook. In fact when it comes to biblical interpretation, the Christian tradition has always recognised that there are various ways of reading Scripture and that the Bible is made up of a number of different types of literature. In short, to quote the words of Galileo Galilei, the central figure in the most famous so-called conflict between science and religion, "the Bible teaches how to go to heaven, not how the heavens go."

What about science? My third point is that science is not a worldview; it's about mechanisms, not meanings.

Science is not a worldview: It's about mechanisms, not meanings

Physics and chemistry do not make claims about the purpose of particles or the meaning of molecules. That's not what they're about, and if we look to science to answer such questions we expect more than it can offer.

Let's have a cup of tea to clarify this difference between a worldview and the pursuit of science. If I point to the kettle and ask, "Why is the water boiling?," the alert physics student will talk about the raised energy levels of the molecules induced by heat from the stove. To which I might reply by dipping my tea bag in a cup; the water is boiling because I want a cup of tea.

Both answers are correct! Why? Because the question, "Why is the water boiling?" is ambiguous. It could be a question about mechanics: "What causes the water to boil?" or it could be about meaning: "What is the purpose of the water boiling?" There are many such questions; "Why are we here?" has both a theological and a scientific answer. "Why is she crying?" has an answer in terms of brain chemistry and neuronal firings that is pastorally unhelpful.

So, we have clarified at least two sorts of questions, which, as shorthand, I am referring to as those about meanings and those about mechanisms. And because science is about mechanisms and not meanings we can see that there are limits to science imposed by the nature of the sorts of questions it answers.

So—my fourth point—science has its limits; in particular science is constrained by the presuppositions it must make in order to do its work.

Science has limits: It relies on presuppositions

As a pursuit of knowledge about the natural world, the natural sciences cannot delve into philosophical or religious questions; these questions are not the subject matter of science. But that doesn't mean that science can leave such issues aside.

The life and breath of science lies in its rigorous approach to uncovering the truth of the natural world based on working assumptions that it does not question. This recognition that science doesn't start from a blank slate, that science must assume some things to even get off the ground, is captured by atheist philosopher Daniel Dennett who warns of the risk of a naïve attitude to science that fails to see its philosophical foundations: "There is no such thing as philosophy-free science; there is only science whose philosophical baggage is taken on board without examination." So, as C. S. Lewis explains in his excellent little book *Miracles*, the philosophical question must come first.

One way of thinking about the philosophical assumptions of science is that they are like tools of the trade that we use to produce results. Think of the carpenter's hammer: it's a tool that the carpenter uses—without questioning it—in order to drive a nail. The focus is on the nail and the hammer is taken for granted.

It's similar with science: science takes for granted its foundational assumptions but it cannot justify them scientifically; they must come first before science begins its work. Here are some examples of foundational assumptions of science.

- *Science can only be practised by assuming that the universe is governed by laws*—there are laws of nature which result in the possibility of repeatable experiments. This means that in the laboratory, the scientist must assume that the results of an experiment are due to the laws of nature and not to supernatural causes. This assumption governs the scientist's methods of going about science and it is an assumption that cannot be proven.
- This regularity or uniformity that science is based on is revealed in the way that *science depends on inductive argument*. Induction is the process of observing repeated events or experience and drawing the conclusion that future events will follow the same pattern. For example, if I observe a million swans and they are all white I might conclude that all swans are white. But as this case shows, induction is not foolproof; Darwin arrived in Australia and found black swans. So science relies on induction but has no way of justifying this confidence.
- *Science must assume that there is a world "out there"* that is independent of what any human being might think or say about it. It is notoriously difficult to prove the existence of the "external world." It is simply something we accept as true and it seems absurd to demand proofs for what we take to be so obviously true without question.
- Finally, *science must assume that human reasoning leads to truth*. Why do we believe that our reasoning and memory and sensory functions are sound? Again, we cannot prove these presuppositions because they are assumptions we must make in order to think about anything.

These are some of the foundational but unprovable beliefs of science. They are all things that science must take for granted in order to get on with its job. In the next chapter we turn to the crucial issue of the relationship of science to a worldview called naturalism, which assumes that science is the only way to truth.

A fuller version of this chapter is published in *God and Science in Classroom and Pulpit*, by Graham Buxton, Chris Mulherin and Mark Worthing. Revised edition, Morning Star Publishing, 2018.

46. Science as ideology betrays its purpose

By Chris Mulherin

This chapter examines "atheist naturalism," the belief that the natural world is all there is, and that scientific knowledge is the only authentic knowledge.

In the previous chapter I suggested that rumours of divorce between science and Christian faith are ill-founded. Faith and science are different but compatible: one is about meanings, the other about mechanisms.

We also saw that science has philosophical limits: there are things that science must assume to even begin its work. Two foundational assumptions that science must make are, first, that the natural world is governed by a regularity that we call "laws," and second, that human logic and reasoning allows us access to the truth about that world.

In this chapter we turn to the crucial issue of the relationship of science to a worldview called Naturalism. In the hands of New Atheists like Richard Dawkins the naturalistic worldview turns into the ideology of *scientism*, which assumes that science is the only way to truth. Wendell Berry, the Christian farmer and writer, warns of the irony inherent in this sort of thinking: "Whatever proposes to invalidate or abolish religion is in fact attempting to put itself in religion's place."

Naturalism is a worldview

Naturalism (with a capital N) is a worldview that holds that there is no God or gods and that the natural world, which science investigates, is all that there is. According to the naturalistic worldview, reality is only made up of "natural" components such as matter and energy. In its cruder forms Naturalism equates Christianity and other faiths to belief in fairies at the bottom of the garden or celestial teapots or the Flying Spaghetti Monster.

Expressed this way we can see that Naturalism is a worldview in competition with other worldviews. It is a belief system that answers the questions of meaning discussed in the previous chapter, although its answers are of the nihilistic variety.

Science and Christian Faith: The State of the Marriage

Science is based on methodological naturalism

At the heart of a good understanding of the relationship between science and faith is the difference between methodological naturalism (small n) and Naturalism (the worldview). *Methodological* naturalism is not a worldview and is an essential foundation of science. It is a tool of the scientific method, a working assumption of science; it is the assumption that when we do science there is no supernatural intervention taking place.

The role of science is quite appropriately to look for natural explanations. This means that supernatural explanations are ruled out in the laboratory and in scientific thinking. Like the carpenter's hammer, methodological naturalism is an instrument used in order to get on with the job. So, although the scientist who uses the tool of methodological naturalism may be a religious believer, their religious belief plays no part in the way they do their experiments.

Now we arrive at a major source of confusion and we are going to have to do a little amateur philosophy.

The success of science does not prove that Naturalism is true

Much of the claimed conflict between science and faith arises from confusing the tool of *methodological* naturalism with a commitment to the worldview of Naturalism. This is particularly evident when people ask a question such as, "But doesn't science disprove religion?"

It seems that what lies behind such thinking is an argument that goes something like this:
Science is based on naturalism
Science is successful
So, naturalism must be true
But Naturalism and Christianity are contradictory worldviews
So, Christianity must be false.

Now there is a major flaw in this argument. There is sleight of hand where the word "naturalism" is used in two different ways (note the lower case "n" and the upper case "N"). We can see this if we rewrite the first part of the argument more clearly as follows:
Science is based on *methodological* naturalism
("God does not intervene in our experiments")
Science is successful
So, *Naturalism* ("there is no God") must be true.

Science as ideology betrays its purpose

But as you can see, the conclusion doesn't follow, because the conclusion talks about Naturalism while the first line talks about *methodological* naturalism which is another thing altogether. In simple terms, just because science assumes that God does not intervene in experiments (*methodological* naturalism), does not mean that God does not exist (Naturalism). So, the success of science can only lead us to conclude this: if God exists, then God normally allows the "laws of nature" to take their course.

So, science seeks truth about the natural world by using the tool of *methodological* naturalism. But science is not committed to Naturalism which is a worldview.

Let's turn now to scientism, which is an extreme version of Naturalism and is an aberration of true science.

Scientism is an aberration of science

We saw in the previous chapter that there are many presuppositions of science that underlie scientific practice but which cannot be arrived at by using science. That is, science cannot show that its own presuppositions are valid. But scientism rides roughshod over these subtleties.

Scientism is a word that is usually used in a derogatory manner to describe a naive, almost blind, faith in science. It is the idea that only scientific knowledge is authentic and any other sort of knowledge is meaningless nonsense.

The thinking behind scientism goes like this: if a naturalistic worldview is correct—that is, if there is no God or gods and the natural world is all that there is—then the only possible knowledge we can have of anything is scientific knowledge. So all that "is" and all that "can be known" is verifiable or falsifiable through science; whatever can't in principle be analysed and measured by science is empty belief and fantasy.

Thus science is held up as the absolute authority in every area of human life and thinking. Instead of science being a tool in the search for truth it has become an ideology—some would say a quasi-religion—that constrains what sort of truths are allowed to exist. New Atheist philosopher Daniel Dennett promotes this view of science when he says, "When it comes to facts, and explanations of facts, science is the only game in town."

What, then, is the problem with scientism? To put it bluntly, scientism is a faith—a blind faith at that—because it boldly claims that science is "the only game in town" or suggests that science can show that Naturalism is correct.

Science and Christian Faith: The State of the Marriage

At the heart of scientism lies a logical contradiction. Scientism claims to be rigorously scientific and says that we should believe something along the following lines: *"The only things you should believe are those things that science shows us to be true."* Let's call that the "S-thesis."

Now read the S-thesis again. A moment's thought reveals the contradiction of scientism. If we are to believe the S-thesis (that is, that we should only believe scientific claims) then why should we believe the S-thesis itself, which is *not* a scientific claim? In fact taking the S-thesis seriously means that we should disbelieve the S-thesis!

In this way scientism seems to be an attempt to lift itself up by the bootstraps, or, to mix the metaphor, the S-thesis shoots itself in the foot. In which case, the appropriate response to Professor Dennett when he says that only science can give us facts, is to ask him simply, "Is that a fact?"

There are many things we believe that are not the result of science. And, as we saw in the previous chapter, there are many presuppositions that science depends on but which science cannot show to be true. If science were the only game in town then it wouldn't even get to first base because the game of science depends on so many "non-scientific" beliefs.

But, as most scientists, religious or otherwise, know well, science is not scientism and scientism does not follow from science: it is one thing to affirm the validity of scientific knowledge but it is another thing to say that all knowledge must be scientific.

Conflict? What conflict?

For most Christians with a healthy respect for natural science it comes as no surprise that science and faith are not locked in mortal combat and destined for divorce. For the Christian, all truth is God's truth and, to use a metaphor borrowed by Christians over the centuries, both the book of God's word and the book of God's works (the creation) reveal something of the creator of all things.

Science and Christianity are neither in conflict nor are they completely independent. In the famous words of Albert Einstein, "science without religion is lame, religion without science is blind." Or, as Cambridge palaeontologist Simon Conway Morris points out, science and Christianity are more like conjoined twins:

> Science without Christianity is actually rudderless, doomed to investigate a universe in ever greater detail, but in a way that Nietzsche would have appreciated, as a completely meaningless exercise. And

Science as ideology betrays its purpose

Christianity without science risks abandoning rationality for a gobbledegook of syncretistic tosh.

To those convinced that a divorce is imminent, let me suggest that the marriage will endure. As someone has said, "the church of God is an anvil that has worn out many hammers" and while the New Testament exhorts Christians to "be prepared to give an answer" to those who ask the reason for their hope, there is, from a Christian perspective, no cause for alarm. The universe is in good hands and they are not those of Professors Dennett or Dawkins.

A fuller version of this chapter is published in *God and Science in Classroom and Pulpit* by Graham Buxton, Chris Mulherin and Mark Worthing (Second edition, Morning Star, 2018).

47. Science is a deeply religious activity

By Tom McLeish

What we now call "science" has its roots in biblical wisdom, and needs to be understood as a "ministry of healing and reconciliation"—and as one of God's greatest gifts and callings.

"Do you know what binds the cluster of the Pleiades ... Do you have wisdom to count the clouds?" asks the voice of God from the whirlwind, within the stunningly beautiful catalogue of many other nature-questions in the Old Testament wisdom book of Job. I have long thought this text to be the tributary from which all biblical explorations, along with our own explorations of the natural world flow, balancing as it does both the light and the dark sides of the world; the sunrise and the hurricane, the known and the unknown.

If discussions of science and religion sometimes get bogged down in Genesis, perhaps that is because they have not made the preparatory journey through the rich material of the Wisdom books. The nature-writing within Wisdom also contains a collection of creation-stories. These use simpler language and metaphor than the developed Genesis texts. They speak of creation as setting in place boundaries and foundations, demarking the heavens and the earth, order and chaos.

The tradition reaches its zenith in the book of Job. Scientists of all faiths and none are invariably impressed by their first reading of "The Lord's Answer" (Job 39–42) with its ancient exploration of the stars, meteorological phenomena, the living world, and strange unknown beasts. Job's complaint is that the creator is as out of control of moral justice as he is of the workings of creation itself.

God's final answer to him has a striking and unusual form: each verse is a probing *question*. Although the consensus of critical scholarship takes this as a sort of divine "put-down" to Job, there is an alternative reading that takes the divine answer in the wider context of the whole book. This reading sees in the answer more of an invitation to think and to observe than a mere rebuke. God's questions direct Job out of himself and into the world around him.

Significantly, they also respect the beautiful "Hymn to Wisdom" of chapter 28, in which the ability of humans, unique among animals, to peer

Science is a deeply religious activity

into the deep structure of the earth, is set alongside the wisdom of God himself, who sees into the world "by weight and measure." Perhaps in a poem of nature questions is to be found the deepest connection to science—we know that our fundamental creative step is to frame the right question, not to jump to the next neat answer.

I have long hoped to take a scientist's personal reading of Job, and other Wisdom texts, as the starting point to make the case for science as a deeply human and ancient activity, embedded in some of the oldest stories told about human desire to understand the natural world. In *Faith and Wisdom in Science* (OUP, 2014) this starting point has inspired a journey towards modern science that visits stories from medieval, patristic, classical and biblical sources along the way.

Writing in the north-east of England I have, for example, found delight in the scientific writings of our local seventh-century scholar, the Venerable Bede. Not only a great early historian (famously the author of *An Ecclesiastical History of the English Speaking Peoples*), Bede sees his calling to expound wisdom as a Christian scholar to include an account of the workings of nature so that people should not be afraid of it, but should understand.

In his account of natural phenomena, *De Natura Rerum*, he even corrects Pliny the Elder's wrong theory of the hydrological cycle, identifies the influence of the moon as the principal cause of the tides, and ventures a natural explanation of earthquakes as subterranean instabilities.

We could put it this way; the book that humanity is writing, whose current chapter we call "science," has many previous chapters that belong to the same story, even though they may have other titles, such as "natural philosophy" or even "wisdom."

Taking such a "long view" of the history of science with its roots in biblical wisdom, I wonder whether much of the current "science and religion" debate operates within a wrong assumption about the *narrative* relation of science and religion. The activity we now call "science" maintains continuity within human culture as old as any story, art or artefact.

A close reading of modern science from the perspective of ancient wisdom tradition unearths a second damaging, hidden assumption—that religion and science are culturally separated not only by time but by the domains in which they apply. Discussion of "non-overlapping magisteria" (NOMA) and its variants, for example, seems inconsistent with the fields themselves. For both science and religion want to talk about everything in the world. A narrative approach, by contrast, is able to develop an approach to science (or in its more ancient form natural philosophy—the "love of wisdom of natural things") that can draw on theological and cultural roots.

The narrative journey of wisdom soon picks up recurring themes that begin to weave a theological background for science. Although the Bible doesn't speak in modern scientific terms, it does reflect over and again on our human relationship with nature—the foundation on which science builds. At each point wonder and responsibility come together—and the meeting is often painful. From the thorns and briars of Genesis 3, to the destructive earthquake and floods of Job, the terrifying deconstruction of creation in Jeremiah's prophecy (ch. 4) and even the groaning of all creation in Romans 8, we are reminded that Bede was right—we do need to mend our relationship with nature.

Following this theme of pain in human confrontation with nature points out one way to develop a theology *of* science (rather than settling for conflict, truce or separation between theology *and* science). In doing so we recognise that both scientific and theological worldviews must be "of" each other, for each must speak about everything that is. Theology must speak of science, not just to it.

A theology of science embraced fully within mission, teaching, worship, prayer and practice also urges the church to drop the view of science as a threat, but rather to see it at the heart of our human calling to live as agents of healing and hope within the natural world. The approach through biblical wisdom in both Old and New Testaments can begin to add colour to what a "theology of science" might mean. There are seven strong ideas that emerge:

- A linear history from creation to new-creation—learning about nature is one aspect of our story that makes the future different from the past. We know more now than we did and less than we will do.
- The astonishing human ability to understand matter—seeing deeply beneath the surface of phenomena is what God does, and calls us to follow.
- The biblical balance of wisdom with knowledge of nature makes us think carefully about the other balance of science with technology—we have a responsibility to work in fruitfulness with the world rather than exploit it.

Science is a deeply religious activity

- The pain of the human-nature relationship reminds us that, like all callings, engaging with nature under God's authority will not be easy—at its simplest level it affirms that doing science is hard.
- The tension between order and chaos is ever present—as well as being the cause of storm and earthquake, it also reminds us that a perfectly ordered, crystalline world is a dead world.
- The central role of questions affirms the risky and open journey—and the humility of living as learners.
- The exercise of love needs to be present—both in dealings with the natural world and among the community of disciples who answer God's invitation to Job and those who follow to seek answers to its deep questions.

Each of these seven themes finds important application with science today and its role in society. For example, the ancient theme of chaos motivates a closer look at the way that random motions of molecules are at work underneath all living processes. Or, to take the role and work of love, this idea of doing science as an expression of love initially appears strange, but is actually an honest experience that might do much to demystify science and reveal the deeply human commitment it draws on.

A condensed statement of this theology of science draws on St Paul's masterly summary to the first-century Corinthian church of the Christian calling to service that he shared with them: "we have the ministry of reconciliation." If the business of those who follow Christ is the healing of broken relationships, as St Paul would have us understand, then when we do science, perhaps we are expressing a "ministry of reconciliation" with nature. Like all damaged relationships which start in ignorance and fear, and lead to hurting both parties, the process of healing is therefore to replace ignorance by understanding, then fear by wisdom, and finally harm by fruitfulness. Surprisingly, science becomes a deeply religious activity—it finds a natural place *within* a religious worldview, not opposed to or threatening belief.

There are urgent lessons for the church from biblical wisdom and a long human story of science: thinking through the purpose of science within the calling of the people of God might equip the community of believers better to deliver a distinctive voice into the troubled public world of science and technology. There are important decisions to make, and make soon, on the political process of decision-making in science and technology: our relationship with the global environment, our ability to manipulate the genetic code. There are better ways of treating science in education and in the media, and healthier narratives by which both religious and secular communities can celebrate and govern science, than those that currently dominate the public forum.

Understanding science to be situated within a larger, biblically-informed, theological project of healing and reconciliation shows that, far from fearing its consequences, the church can embrace it as one of God's greatest gifts and callings.

> The same ideas found in McLeish's *Faith and Wisdom in Science*, expressed through personal story for a wider readership and especially aimed at high school students, have been published in a book that Tom McLeish has co-authored with science-teacher Dave Hutchings: *Let There Be Science*.

48. Science cannot be an exclusive guide to reality

By John Pilbrow

Science is not sufficient to explain all dimensions of existence, and to see science as our only valid guide to understanding reality is a great mistake, argues British philosopher Roger Trigg in a 2015 book.

It is in the context of the increasingly scientific and technological world we inhabit that New Atheists such as Richard Dawkins bring their claim that science provides the only sure path to knowledge. This point of view, referred to as scientism (or scientific materialism), actually restricts reality to what "is within reach of the sciences." Another voice is postmodernism, which asserts that "truth is not encountered but constructed with a consensus" that may actually be culturally and politically dependent. Neither of these perspectives leaves room for any idea of a universal truth.

To rebut the attacks on reason in science, particularly from scientism and postmodernism, Professor Roger Trigg has contributed a timely and very readable monograph, entitled *Beyond Matter: Why Science Needs Metaphysics* (Templeton Press, 2015), about the relationship between science and reality. While the author affirms that science discovers truth about the natural world, he asserts most strongly that it cannot explain everything.

This is a book I hoped somebody would write someday. The author needed to be a recognised philosopher with a deep interest in modern science and its roots and history. And if that philosopher were also a Christian believer, better still. Roger Trigg fulfils these criteria. His book is not primarily about a Christian view of modern science. Rather it is aimed at explaining to a wide audience both inside and outside the scientific community that science does not exist on its own.

While recognising that the term metaphysics is likely to be unfamiliar to many readers, Trigg's straightforward definition that metaphysics is a branch of philosophy about reasoning from first principles should help. It is the task of metaphysics to identify the prior assumptions (metaphysical assumptions or presuppositions) that are required to pursue scientific investigations. These metaphysical assumptions, while essential to the practice of science, are not derivable from within science, contrary to what people like Richard Dawkins believe. Trigg's point is that science is not self-authenticating.

Metaphysical assumptions in science

The following list, gleaned from Trigg's book, while not necessarily complete, is consistent with what scientists actually assume either explicitly or implicitly in order to pursue scientific investigations. They are grouped according to my own preference.

- The universe is rational and ordered, hospitable to human reason and amenable to empirical enquiry.
- We have a capacity to understand the physical world by means of experiments and/or through the making of observations. This involves our human rationality and the freedom to exercise it using the normal rules of logic and sound reasoning.
- Since the physical world does not have to be the way it is, it is said to be contingent and modern science is said to depend on "contingent order."
- Science explores what is the case and it is carried out in the natural world without reference to "supernatural" agency. Put another way, God is not invoked as an explanation for what is not known.
- Judgement is required when weighing up different possibilities when developing theories to explain existing phenomena and that also may predict new phenomena.
- The physical reality science investigates is distinct from us and exists independently of whether we observe it or not.
- This is known as critical realism, which is the attempt to maximise the correlation between what we know and the underlying reality behind it. There is good reason to believe that scientists are critical realists.
- A third-person stance is adopted to ensure objectivity, ensuring that the experimenter is distinct from what is being investigated.
- Since scientific results may be replicated anywhere in the world, the practice of science is said to be universal.
- Laws of nature that describe observable patterns are descriptive, not prescriptive.

Popular ideas from the twentieth century

During the past century, science was strongly influenced by several dominant ideas attributable to both the "Vienna Circle" and Karl Popper.

The Vienna Circle was a group of philosophers who regularly met in Vienna between the two world wars. There is not space to outline their

Science cannot be an exclusive guide to reality

views in detail. However, amongst other things, they claimed that anything science could not explain should be regarded as nonsense, leaving no room for metaphysics. That is pretty much what people like Richard Dawkins and other New Atheists believe. Confronted by the new quantum mechanics in the 1920's, members of the Vienna Circle argued that truth about the universe is only actualised when an observer makes a measurement. This point of view has been criticised as simply dressing up the old idea that a clock only ticks if it is heard. In the main, these views have little traction today for scientists investigating "the nature of a physical reality that exists independently of human understanding."

It was just after World War II that the famous philosopher Karl Popper introduced the principle of falsification in science. He believed that scientific theories could never be said to be proven, but they could be demonstrably falsified by new evidence. From my own experience, while the principle of falsification is sometimes of value, its rather negative connotation obscures the major task in science which is to come up with meaningful explanations.

Turning to postmodernism, attempts to apply its ideas to science falter in face of the high degree of agreement amongst scientists about the "big picture" theories such as big bang cosmology and evolutionary biology. Such level of agreement transcends nationality, culture and religious affiliation or none.

Scientism does not fare any better with its claim that being scientific means "trusting science as our exclusive guide to reality." It is logically inconsistent since it denigrates "the free use of human reason at the same time as upholding science as the greatest expression of that reason." Both scientism and postmodernism face an awkward philosophical dilemma. "Why practise science at all, if there is a risk that all is illusion or fantasy?"

Deeper issues

Trigg argues persuasively that none of the ideas discussed in the previous section settle the fundamental question as to whether there is an independent reality or why empirical science could be the only path to human knowledge.

We seem to be left with a stark choice. Either we must simply accept the order in the physical world and the way things are as brute facts, or, alternatively, trace the fact of order and regularity back to the purpose of a creator. A good recent example of such a choice concerns the fine-tuning of the universe for life as revealed by modern physics (called the anthropic principle). I find it rather surprising that it was the former Astronomer

Royal, Martin Rees, though not in fact a believer, who suggested that fine tuning might turn out to require a theological explanation

Trigg provides an insightful analysis regarding the emergence of complex systems. In so doing, he joins the chorus of those who argue that each level of complexity leads to new properties that cannot be completely explained in terms of the constituent parts.

For example, physics makes use of mathematics but cannot be simply reduced to mathematics. Biology certainly involves physical and chemical processes yet it cannot be reduced to physics. Trigg appeals to metaphysics to settle the matter, saying "Arguments about the reality of a whole involving more than the sum of its parts must be metaphysical."

By way of further example, clearly all living cells have physical and chemical processes going on in them, but the overall properties and behaviour of even a single cell cannot simply be reduced to physics or chemistry. A familiar example comes from blood tests. The results include levels of cholesterol, triglycerides, proteins and much besides. However, while this information tells us something important about ourselves and our state of health, in no way is it able to describe us or our behaviour as whole persons.

To reinforce his support for empirical science, Trigg appeals to *The Singular Universe and the Reality of Time* by philosopher Roberto Unger and physicist Leo Smolin. Unger and Smolin argue, correctly, that "science is corrupted when it abandons the discipline of empirical validation." In the light of that, it is not surprising that they object to the idea of multiple universes (or multiverses) in cosmology. They state that "The conjecture of such universes" amounts to a "vain metaphysical fantasy disguised as science." This last comment makes the point that scientists can easily present metaphysical arguments as if they are scientific ones, without realising that is what they are doing! To extricate ourselves from such a problem, we have to realise that the only universe we can investigate with the empirical methods of science is the one we inhabit, not ones we might speculate about.

In what may seem to be a backflip, Unger and Smolin admit, however, that we may "have good reason to accept the existence of many entities" or new phenomena predicted by theory "that lie beyond the reach of contemporary science." Gravitational waves and the Higgs particle come to mind, both predicted long before their existence could be confirmed.

Conclusion

Trigg's main argument is that without a metaphysical framework to support it, the practice of empirical science cannot lead to a reliable understanding of the universe. His recurrent theme is that the universal claims for science which "have to be metaphysical in nature" cannot be rationally justified from within science itself. For that reason, he says the "distinction between science and metaphysics should not be broken down." On this basis Trigg provides solid justification for the pursuit of science as a legitimate and important human activity. The essence of the book is encapsulated in the last two sentences. "Science without metaphysics flounders, as if lost in a vast and featureless ocean. It loses all sense of direction and purpose."

I found *Beyond Matter: Why Science Needs Metaphysics* to be a most engaging book. I like to think it will be read with profit by practising scientists, postgraduates and even undergraduates in the sciences, as well as by those with some familiarity with the philosophy of science. Even though, as Trigg points out, we were not born to think scientifically, there is also something there for those with little or no scientific background, though I should warn that it is likely to prove a quite demanding read! Though this is not a book for everyone, its take home message that science is not self-authenticating is important for all of us to be aware of.

49. Science is "entirely within God's purposes"

By Chris Mulherin

Tom McLeish is a convinced believer in the God-given harmony between science and Christian faith. Professor of Physics at Durham University in the UK, a Fellow of the Royal Society, and author of *Faith and Wisdom in Science*, he believes Christians are called by God to engage in vigorous scientific investigation of the natural world of God's creation, and that science is part of healing creation. He was brought to Australia by ISCAST—Christians in Science and Technology and gave the inaugural annual Allan Day Memorial Lecture at Ridley College in September 2015. He conversed with Chris Mulherin.

Chris Mulherin: Tom, you're a physicist; I believe you're into weird substances, is that right?

Tom McLeish: My area is soft matter physics, which is more or less what it says on the tin—it usually comes in a tin—soft, gooey stuff: jellies, polymers, pastes ... the ways that the molecular structure—the tiny building blocks of the plastic matter—give rise to the properties of fluids, the stickiness, the gooeyness, the sliminess.

Things aren't what they seem

Talking about molecular structure, in your lecture at Ridley College, you suggested that the stable, hard sort of surfaces of tables and chairs and buildings and bricks aren't quite what they seem. Is that right?

That's right. I'm fascinated by the way that matter at the small scale—when you dive down a billion times into the scale of atoms and molecules—looks very different to how it does to our eyes. We're familiar with the fact that when we sit on chairs we don't fall through. But our understanding of atomic structure tells us that nearly all the mass of an atom is in a tiny little particle called the nucleus, right in the middle, electrons whizzing around the outside; most of the atom is empty space. My school physics teacher used to use an analogy of an atom like a fly in a cathedral. So if we're made of mostly empty space, why when we sit on these chairs don't we fall straight through?

Science is "entirely within God's purposes"

That's a good question; tell us about the role of questions in science.

I think that the way we teach science or introduce science, sadly, puts a number of people off. We get the impression that science is all about getting the right answers. You know: this is how you succeed in school; you get problems and you work them out and you either get the answers right or wrong, and if you get the answers right you get more marks. Actually, it's coming up with the creative questions that's the real challenge. That's the imaginative leap in science: it's formulating the really powerful, interesting questions.

I sometimes have to say to a brand new PhD student who comes to Durham, beginning their career as a research scientist, "I'm sorry, you've been taught that coming up with the answers is the key thing in science, but from now on it's inventing the really creative questions." Like the question, "Why don't we fall through these seats while we sit on them?"

So we're talking about misconceptions of science: what other misconceptions are there?

We sometimes hear a lot written about scientific proof, but I'm happy to admit, there's no such thing as scientific proof. No science ever proves things; we know things with greater or lesser degree of certainty, but we never know anything for absolute sure.

I think one of the other misconceptions people have with science is that science is somehow antithetical to or irreconcilable with religious belief—Christian belief, for example.

So you're not enthusiastic about the idea of science versus faith. What about speaking of science and faith: two distinct areas of human endeavour that perhaps don't conflict but nevertheless are distinct?

I think that they're closer than that. I've thought a lot about the relationships between a Christian worldview of a God that makes us and the world. I've been a Christian since my young adult life. I've had some theological training as well as scientific training, and I'm also a lay preacher in the Anglican Church. It has always seemed to me to make a lot of sense of a universe full of meaning, full of structure, full of order, that our minds can understand.

But in terms of the relationship with science, what does science look like if you embrace it with a Christian worldview? Rather than think of it as something that you just have to be somehow comfortable and compatible with, let's think of it as something entirely within God's purposes.

It's also worth thinking about how science arises, as it indeed does, from a religious worldview. All historians will agree that science arose from communities of religious belief. The early modern scientists like Francis Bacon, Boyle, Newton, and so forth, had explicit theologies that what they were doing in creating experimental science was in obedience to God's commands to recover lost knowledge of the world through the fall. This is just one example of how science has emerged from faith historically.

A theology of science

So what is your theology of science, in a nutshell?

Well, it comes from two types of biblical reading, one in that wonderful corpus of literature, the Old Testament wisdom literature; I think I can trace the early roots of science back to what in the ancient world was called wisdom. Reading Old Testament wisdom literature opens up the human relationship with the natural world as being something that God calls us to develop and to look on the natural world with the same love and care that he does.

What books are you thinking of?

Proverbs, the Psalms, and supremely, the shining, snowy-mountain peak of wisdom literature, which tells us about our relationship with nature, is the book of Job.

Not Genesis 1 and 2?

No, no. They come later. There are good creation stories in Genesis 1 and 2, but they're very formal, they're quite difficult, they're not "Creation 101" if you like. They're important, but their time of writing is not particularly early and they're not particularly fundamental. There are creation stories all through the Bible: last time I counted I got to about thirty before I gave up counting.

So, a theology of science for me, with all this rich biblical material throughout Scripture in mind, is the story of making friends again with creation. Paul talks about our job as Christians as the "ministry of reconciliation." You might say that for Paul, being a Christian is about healing broken relationships. Everyone gets that; you don't have to know deep theology or Old Testament to know that mending broken relationships is something we all need and the world needs badly.

One of those broken relationships is between people, humans and the material world. We don't understand it, we don't know about it, we don't treat it very well, we're frightened of the way it treats us (storms, earthquakes, tsunamis, all that stuff).

Where does climate change fit in for the Christian?
Well, one thing is clear: humans do have a responsibility for stewarding nature, which means we really do have the power—biblically we have the power—to steward it for ill or for good.

The myths of conflict

Can I ask you about the aggressive so-called New Atheism? Out in the street, and particularly in the public press, there's a war going on with Christian faith.
Well, the press loves conflict, and conflict leads to publicity. So it's very convenient to promote a conflict myth. But there is no conflict between a life of faith and science. Far from it. Science historically, philosophically, personally and functionally can flow and should flow from faith. But you know, keeping alive a conflict narrative is a personal agenda for some people; I'd like to defuse that conflict, because I think it does huge damage to science because it makes dehumanising claims about science.

There's another side of the conflict myth too, isn't there, from the Christian point of view? That is to say, there are some Christians who think we are in fundamental conflict with science.
Yes, you can create a conflict with science; six-day creation—the belief in a literal six-days—conflicts with the history of thousands of years of biblical interpretation: the ancient Jews and Hebrews never believed that. It is a largely modern social phenomenon and doesn't have anything to do with authentic, orthodox Christianity through the ages. It's perversely pushing the Bible into being literature that it never was, and it's trying to make a lazy shortcut from God's command to do difficult and careful science into thinking that he's given us the answers in the back of the book—or in this case, in the front of the book. But it's unnecessary, it's damaging, and it's doing a huge amount of damage to the church and to everybody—it's confusing children. I think we've got to stop it.

Before we finish ... you've just written a book, haven't you?
Yes, that's right. It's called *Faith and Wisdom in Science*. I wrote it because I wanted to swim in the ocean of ideas that you swim in when you drop the head-on conflict narrative between faith and science and you say, "Well,

just suppose you take the view that science is something entirely within God's kingdom," and you start asking what it's for. You start looking at the Bible in terms of our relationship with nature; you look at the history of scientific thought and you explore the relationship between science and the other humanities. So I wanted to bring all that together and ask whether it's possible to reframe science in a much longer, much more human story that sits entirely within the Bible story of creation, fall, redemption, and the future hope of new creation.

A full version of this interview is available on the ISCAST website at: http://iscast.org/resources/faith_hope_love_science_McLeish_interview

50. Science is wonderful, but it's not the only game in town

By Robert Martin and Chris Mulherin

New Atheist leader Richard Dawkins, who was to be in Melbourne, Australia in March 2016, was invited to dialogue with the Rev. Dr Chris Mulherin on the City Bible Forum's "Logos Live" radio program, but Professor Dawkins fell ill, cancelling his trip. However, the program, on the question "Can we live by science alone?" went ahead, and Chris Mulherin was interviewed by Rob Martin, Director of CBF in Melbourne. This is an edited version of the interview.

Robert Martin: Chris, what made you interested in the relationship between science and Christianity?
Chris Mulherin: I became a Christian at university; I moved from engineering to philosophy and then theology and became very interested in the nature of science—how science works, is there a scientific method, those sorts of questions—and then more in the relationship between science and Christian faith.

What was it that made you particularly interested in exploring that further?
I think it was an apologetic interest, because although I never found a conflict—and many, many Christians who are scientists don't—it's quite obvious that lots of people seem to think there is a fundamental conflict between science and faith: Richard Dawkins is the most prominent person promoting that view. I think it's very important for Christians and the Christian church to realise that there is no conflict, and for people who aren't Christians to realise that believing in mainstream science should not be a stumbling block to considering Christianity.

Well indeed, Richard Dawkins, possibly the world's most famous atheist, claims that religion and science are locked in mortal conflict. Another prominent atheist, Daniel Dennett, says "When it comes to facts and explanations of facts, science is the only game in town." Is he right?
Well, of course he's not; science is not the only game in town because science can't give us answers to what human beings mostly call the most important questions of life.

Such as ...

The most important questions of life are about meaning and purpose, not about mechanisms and particles—science is brilliant at giving us answers about mechanisms and particles; science is exciting; science reveals truth about the world we live in; science is a wonderful thing. But it has its areas of expertise, and its areas of expertise do not tell us why we are here, whether love matters; it doesn't tell us about morality, purpose, meaning ...

So there are some limits to science and what science can do. So, maybe before we go any further, what exactly is science? What do we mean when we talk about "science"?

Well, that's a good question because many people just assume they know. We've been using the singular of the word "science" ... There is no "science"; there are "sciences" and to a greater or lesser extent they have family resemblances.

But couldn't we adopt what is known as the scientific method? In *A Devil's Chaplain* **Richard Dawkins explains to his daughter that "scientists, the specialists in discovering what is true about the world and the universe, work like detectives: they make a guess, called a hypothesis, about what might be true. Then they say to themselves, 'if that were really true we ought to see so-and-so': this is called a prediction." Dawkins then outlines how we test predictions and the result is evidence, and evidence is good reason for believing something. So, isn't the scientific method a good and reliable way of knowing what's true and right in the world?**

That sounds very nice, and in some areas of science that works better than others. But there's a lot more that most people would think was true and right in the world that goes beyond that. For example, did Julius Caesar exist? Well, let's work out what hypothesis leads to a prediction that we can test. But there's no prediction that we can test for the existence of Caesar; does that mean that the question of Julius Caesar's existence is not either true or false? No, it just means that it's different: we access the truth of the existence of Caesar in a different sort of way to the way we access the truth of whether this is a certain sort of cell or not, and whether it's going to do this or that if we grow it in a test tube.

Let's go to a more extreme case: morality. Is it true and right that it's wrong to torture innocent three-year-olds? Most of us would say yes, of course it's wrong to torture three-year-olds. Well let's work out a hypothesis

Science is wonderful, but it's not the only game in town

and a prediction that will prove that ... but that's just not the way you do morality.

So what other games are in town then, when it comes to working out facts and what's true about the world?
Well, I'd like to turn the question around and question the whole nature of science. In the model of science that we've received, we have this idea of objectivity that virtually rules out human judgement. But every scientist knows that there are judgement calls involved in science—Einstein at sixteen years old makes a judgement call about light and relativity and the universe; he is convinced that things are a certain way, and others are convinced they're not. And in the long run, his judgement call was proven right.

So, science is a very human pursuit, which doesn't mean that it's not about the truth of the real world, but every single scientific judgement ever made in history has been made by a human being who effectively says, "This is what I believe to be true."

An example: In these decades we see a vigorous debate going on about climate change. The experts on climate change in the world have made a judgement call that human beings are causing climate change. Their judgement call is based on their deep experience, their lots of research, their lots of conversations with each other and reading academic articles—they are the experts, and the expert judgement on climate change is that human beings are causing it. Other people—mostly who have no idea about climate change—are saying, "Well, no, I don't think we're causing climate change"; they've made their judgement call.

But there are two different sorts of judgement calls going on there: one comes out of the expertise of those who are immersed in the field, and ought to be paid attention to—although I can't *prove* that either—and one comes from a group of people who are mostly not climate scientists. Now maybe there's some little room for argument, but the fact is the experts in the field are basically agreed about the big picture.

So then how do we gain reliable information about the world?
I've given you the example of climate change, but we could take any of "the sciences"—we become experts; we talk to each other; we make judgements; we test each other's theories; we show each other why this theory doesn't work and this one does; and our belief is that science progresses in

that sense; it progresses towards greater and greater confidence about what the truth is. But that's not proof that we are 100 percent sure.

So what makes good evidence then?
It depends on the field. Good evidence for the fact that my wife loves me, or that so-and-so has got NK-cell lymphoma, or that God exists—they're all different sorts of evidence because evidence has to accommodate to the field you're dealing with. Evidence in morality is very different from evidence in physics.

So should we live by evidence?
It depends what you mean by the word "evidence": if by the word you mean what is normally meant, which is "evidence" in the natural sciences, no, we have to go by a lot more than that; and my wife would expect me to go by a lot more than that when she says "I love you; please believe I love you." She would not expect me to get out test tubes and to say, "Okay, there's a good hypothesis; I am now going to set up a whole series of tests, predictions, and see how you perform." I'm not sure that that would be good for our relationship.

One of Jesus' disciples, Thomas, was sceptical of the claims that Jesus had been raised from the dead and he demands evidence. Could you say that Thomas' demand for evidence is somewhat scientific?
I would say it's more existential ...

What do you mean by existential?
Well, it's a cry from the heart rather than a cry from the mind. Thomas has lost his Lord, he's lost the one that he followed; and he's extraordinarily confused about what's going on. These disciples had put their hope in Jesus and he had let them down badly by being crucified on a cross outside of Jerusalem; Thomas is in shock and disbelief.

So, while his expression can be portrayed as scientific—you know, "I need to see the evidence"—actually it's a cry from the heart, I think. And Jesus responds to the cry from the heart by saying, a week later, "Here I am; come and put your fingers in my side and you'll see it's really me."

But I'm not sure it's a model of the scientific method.

I think Jesus presented *himself* ... and knowing Thomas well he knew the word that Thomas needed, which was, "Get a grip, Thomas; it's really me. Let's move on." And it's interesting that he says, "Stop doubting and believe." I think there are two sorts of doubt: there's the sort of honest, wrestling doubt that seeks the truth; and there is the doubt that really is not interested in the truth if it happens to lie in the direction of Jesus, and therefore it's more appropriately called scepticism, a sort of chronic scepticism.

Or hyper-scepticism?

Hyper-scepticism that will keep on asking the question, "Well prove it; can you prove it?" And that just closes down all moving forward, all progress, in terms of opening yourself up to the truth, wherever it lies.

So what would you say, then, to the hyper-sceptic—someone who says, "Prove it, prove it!?"

Well, I'd talk about climate change; I'd talk about the love they have for their spouse or partner; I'd talk about areas of life where they don't have that sort of proof. And if it were Richard Dawkins ...

We did invite Professor Dawkins to join us ... What words do you have for him? Can we live by science alone?

Well, if I had a private word with Professor Dawkins I would ask him why he's so angry at religious people; I would tease him out a little bit. Because I suspect I would probably agree with a lot of the things he thinks about the evils of religion, or the importance of his relationship with his daughter.

I would like to convince him that actually most of the religious people I know are not like the caricature of the religious people that he attacks. I would like to try and open a door to him to think that, firstly, to be religious isn't necessarily to be like the stereotype that he has, and secondly, that science is not the only game in town.

51. There is more science–faith overlap than ever

By Tom Butler

Our world is in crisis and people of faith must work with scientists for a world of hope rather than despair.

I have inhabited the worlds of both science and religious faith during much of my ordained ministry, particularly when I served as chaplain and lecturer in electronics: first at the University of Zambia, and then at the University of Kent. One of the projects I worked on in Kent caught up with me in later life.

I helped a group in the electronics department with a specialised radio to fit in a tank, which I discovered was destined for Iraq. This was when Saddam Hussein was being supported by the West in his war with Iran.

Some 20 years later, as Bishop of Leicester, I was giving medals from the first Gulf War against Iraq to some soldiers at an intelligence and communications base in Leicestershire. At the end of the visit, a young captain showed me some of the equipment that they had captured in the war.

To my shock, it included the radio I had worked on. I said: "I've got a confession to make—I helped develop that radio."

"Don't worry, Bishop," he replied. "We know, and we've got a copy of your circuit diagram."

Throughout my ministry, the worlds of theology and science have overlapped, and my attitude to both has changed. If, when I was studying and teaching science, I managed to snatch opportunities for reading theology with almost a secret thrill, since becoming archdeacon and then bishop, I have found my position rather reversed: now, the thrill is picking up the latest science magazine or paperback. For me, God is as likely to have an impact on me through either, and in particular in the overlap of these worlds.

I have a notice above my desk, which for me goes to the heart of the overlap. It is from a postcard sent in the mid-nineteenth century from one scientist, Maxwell, to another, Joule, after Joule had made a breakthrough in his research. Maxwell wrote: "There are very few people who have stood where you stand and after a period of minute observation and patient mental toil have put their minds into exact accordance with things as they really are."

There is more science-faith overlap than ever

To put your mind into exact accordance with things as they really are—that is the objective of all scientific research. To live in accordance with things as they ultimately are is the objective of the life of religious faith; for faith in God is not the same as having a hypothesis about God. In faith, you are the apparatus, and your life is the experiment.

Albert Einstein famously wrote: "Science without religion is lame; religion without science is blind." It seems to me that the world is in crisis, and in such a world it is essential that people of faith and of no faith work together with scientists for a future of hope rather than despair.

As a Christian, I don't have to believe in the literal truth of the Genesis story of creation; for the book of Genesis is not a science, history, or geography book. I accept that the theories of big bang creation and the evolution of life give us a great insight into the creation of humanity, and in their own terms are just as inspiring as the Genesis story.

But that does not mean that the book of Genesis does not also contain profound truth. It is the fruit of the reflection of the people of Israel on the triumphs and disasters of their history. If the big bang theory postulates the "how" of creation, Genesis attempts to answer "why."

There is nothing particularly new in this insight. Galileo, back in the fifth century, wrote: "The Bible does not teach us how the heavens go: it teaches us how to go to heaven."

It is instructive to compare and contrast the scientific and biblical stories of creation, however, for I believe that each has much to learn from the other, and each can add value to the other.

The big bang creation sees the big bang at the start of time and space, some 13.5 billion years ago, creating in a big bang something out of nothing. The something was very strange indeed: a big bang exploding from a pinprick of searing heat and incredible density, until the laws of physics kick in after 10^{35} seconds, exponential inflation occurs, and the building blocks of the simplest atomic nucleus coalesce. We see an ordered creation out of nothing, but following the laws of physics over both incredibly small and then incredibly large periods of time.

The story as told by Genesis also sees creation out of nothing; yet it is no accident, but a purposeful creation because God wills it. Before there is anything, we are told that the Spirit of God brooded over the dark void, rather like the laws of physics brooding over the primeval big bang chaos.

Science and Christian Faith: The State of the Marriage

It is again an ordered creation in the writer's mind over a period of six days, and the procession is remarkably similar to the evolution of the galaxies, and then life forms, as in the scientific theory.

Yet, at the end of each day, the creator reflects and says: "That's good." And in the sixth day, humanity is created "in our own image"—to share something of the mind of the creator and his joy in his creation. "That's very good."

In the time-scale of creation, we are very recent kids on the block, and we inhabit a very special world where life is extremely fragile, and could easily be extinguished through our own stupidity. Science and religion need to be partners in avoiding that calamity.

If science strives to tell the "how" of creation, and religion the "why," I believe we need both, and can have both. There is nothing in the big bang and Darwinian theories that gets in the way of my religious belief in a transcendent God, who out of pure goodness and will created space and time out of nothing. As part of the package, God created the laws of physics, which then made creation make itself.

I would claim that science has developed in the West because it has been built on the foundations of the Christian religious principles of the value of the created world and the importance of rigorous thought.

Yet we religious believers do have to pay attention to the arguments of our critics; for the average person's view is likely to be that science has "disproved" religious dogma through its rigorous discipline of testing hypotheses by experimentation.

In this way, since the Enlightenment, science has been dramatically successful in extending knowledge and understanding of the universe. Theology, the "Queen of the Sciences" of past ages, is now tolerated in religious belief and practice as a private preference, but in no way has the authority of the true sciences.

This may still be the position of the person in the street, and is articulated clearly and with conviction by Professor Richard Dawkins and others, but I believe that the situation is becoming less certain, even though the person in the street may not catch up with this thinking for some decades yet.

Since the beginning of the twentieth century, traditional science has been challenged by a "New Physics," which describes a world that is very strange indeed. Sir Arthur Eddington put it this way: "It has been the task of science to discover that things are very different from what they seem."

There is more science–faith overlap than ever

I believe that it is significant that Professor Dawkins is a biologist, and biology has been one of the most successful of the "old sciences" and its children. But it is the discipline of physics that tries to delve into more fundamental levels. It is the New Physics, delving through quantum mechanics into the incredibly small, and through relativity into the incredibly large, which has been discovering that existence is more mysterious than we imagine, or can imagine.

Niels Bohr, one of its exponents, puts it this way: "Those who are not shocked when they first come across quantum theory cannot possibly have understood it."

The quantum world really is different, and the only way to come to grips with it is to suspend some of the beliefs and the disbeliefs of traditional science. It is as though the edifice of science and technology has been carefully and logically constructed, brick by brick, to produce the most impressive structures, only to discover that the foundation on which all this is built is shaky.

The writer Annie Dillard has put it this way:

> For some reason, it has not yet trickled down to the man in the street that some physicists now are a bunch of wild-eyed, raving mystics. For they have perfected their instruments and methods just enough to whisk away the crucial veil, and what stands revealed is the Cheshire cat's grin.

I feel like this when I read about what physicists now understand about the structure of the atom. When I was doing my sixth-form physics, it was relatively easy. The atom, we were taught, was a kind of microscopic solar system, with a core of protons and neutrons, and circulating electrons.

Today, however, we hear of quarks and muons and tauons, of mesons, gluons and gravitons; we hear of elementary particles and anti-particles. Now we don't have to worry only about mass and charge, we have to consider in addition spin and quantum numbers. These elementary particles are now thought to be different vibrational excitations of strings of energy.

Yet the experimental evidence for any of this is slight, or even non-existent. This is why the series of experiments going on now in the large Hadron Collider in Geneva is as exciting as humankind's first setting foot on another planet. We could really be on the cusp of new discoveries that will blow our understanding of the sub-atomic world apart.

Science and Christian Faith: The State of the Marriage

The New Physics does not share the hostility of some traditional scientists to religious faith. On the contrary, it often seems to speak our language, or at least some of the language of Christian mystics.

As the seventeenth-century German mystic Angelus Silesius puts it: "I know God's countersign. His signature is writ in every creature, canst thou but interpret it."

This chapter first appeared as an article in the 7 January 2011 edition of the UK *Church Times*. It is an edited extract from Dr Tom Butler's Company of Educators Franklin Lecture, in the UK. It is available in full at www.churchtimes.co.UK.

Design, the Universe, Fine Tuning, Reality and Metaphysics

52. A purposeful universe

By Paul Richardson

Paul Davies, author of *The Mind of God* and 1994 winner of the prestigious Templeton Prize for Progress in Religion spoke with Paul Richardson in 1997.

Paul Richardson: You have written that science can provide a surer path to God than religion.
Paul Davies: That's often quoted but I wrote it to be deliberately mischievous.

But do you stand by it?
Yes.

One of the points you have made is that you are impressed that the universe seems to follow basic laws, and these laws seem to enable the emergence of conscious life. In explaining this apparent design you have used the word "God" but you have made it clear that you don't want to use that in the Christian sense.
I like to distinguish between the concept of God as the grand architect of the cosmos, a sort of remote, timeless, intellectual underpinning of existence, and the personal God of popular religion—the God you might pray to, the God who might work miracles on your behalf, the God who would intervene on a day-to-day basis in human affairs. It is very hard for a scientist to believe in the latter type of God, but when it comes to the former type of God—the notion that the universe is not absurd, that at rock bottom it is deeply rational and that there is something like mind or intellectual input underpinning it all—that's something that appeals to a lot of scientists, or certainly a lot of physicists, and something about which there is a lot of circumstantial evidence from scientific research.

So what I usually say is that the God that I am referring to is this grand architect notion of God, and I might say that many eminent Christian theologians over the centuries have also, when they have spoken of God, had that in mind.

When it comes to the notion of a personal God I suppose I am agnostic on that, but I see no reason why science can't help us to illuminate it.

I feel deeply uncomfortable with the notion of the incarnation. I think that puts me where I can't be a Christian, although we know of course there are some senior members of the clergy who are equally uncomfortable but would still call themselves Christian.

In your writings your concept of God sounds rather vague and nebulous. It is rather like seeing smoke and saying "I see that which causes smoke." It would be more useful to say "there is a fire."

I believe that just giving a label to that which we don't know tells us nothing. Atheists are well used to the argument that simply saying "God has brought this about" doesn't have any value unless you explain how God brings this about.

I think it's better to ask the question: "Is the universe ultimately absurd, is it about nothing at all or is there something like meaning or purpose in the universe?"

The words you use—purpose, design and so on—are intentional words which we associate with personal being. Doesn't this suggest that what you are really talking about is a God who is in some sense personal?

Well yes, it could be interpreted that way. Certainly one way of interpreting the ingenuity, harmony, felicity and apparently contrived nature of the physical universe is to assume that this is simply designed by a being that we would recognise as a person with a powerful intellect, a loving being, a rational being … and that this being has somehow conjured this lot into existence. You could do that.

There are a number of reasons why I wouldn't want to reduce it to quite such an anthropomorphic level. One of the reasons concerns the nature of time which is a real problem for physicists. Physicists are very used to this idea that time is part of the physical universe, and any sort of rational scheme that is going to explain the physical universe has to be something that is ultimately outside of time. And I'm not sure that anything you would recognise as a person or personal type of being can be truly timeless. Now some theologians, for example Keith Ward, attempt to fuse the temporal and atemporal in a single complex entity. Maybe that's one way to go.

But it seems to me that there is a danger of getting back to the miracle working cosmic magician who brings about this felicitous and harmonious state of affairs by continually prodding and intervening with nature, moving atoms about fitfully, as John Polkinghorne would say, and I think that's a horrible image.

A purposeful universe

What I have in mind is something much deeper and more subtle, and in many ways more impressive, which is that if you have got a cleverly thought out—so to speak—set of physical laws; then all of this happens automatically. The universe brings itself into existence, it organises itself, life emerges, consciousness emerges ... all of this is part of a natural outworking of this felicitous set of laws.

If there is some sort of God, some architect of the universe, who has arranged the laws so that the universe will exist ultimately without the kind of intervention that you don't like and will produce conscious human life, don't you think it strange that this "God" doesn't seem to want to make contact with the conscious human life, that there is no sort of disclosure, no sort of revelation?

If you really don't want an interventionist God, then even mental disclosure would amount to physical intervention. If we believe that our thoughts are tied to patterns of activity in our brains, then to put it crudely, to plant a thought in our minds would amount to moving atoms about in our brains. That notion of God as just another force of nature is I think a rather demeaning view of God.

Theologians have suggested various ways God intervenes but I'd like to push a little bit the question: What's the point of God arranging the laws so that conscious human life emerges and then not entering into any relationship with this human life?

What is the point? Well now, if I knew the point of the universe I think I would have told you already. I could equally turn the question around and say: If God is perfectly capable of bringing about a state of affairs by that sort of intervention, what is the point of going through the whole drama of the evolution of the universe in any case? He could simply have created the whole thing in its final form.

I think it was designed to preserve human freedom ... Is there a danger that some scientists are themselves becoming deified?

I am well aware of this, and of course it would be quite wrong to resort to a sort of scientism where science replaces religion or science becomes a religion, and scientists claim to be talking about truth with a capital "T." I think John Templeton is spot on in his belief that if science is done properly it should never claim to be dealing in truth. I think the best description of science is John Ziman's, which is that it is reliable knowledge. That doesn't mean it's infallible.

The great strength of science is that we are always dumping the old theories and replacing them by new. That doesn't mean the old theories are valueless; it doesn't mean that they're wrong; it's not a matter of right and wrong. You know Newton's theory of mechanics is used for most purposes—you can get a spacecraft to the moon quite well, but we've got a better theory—we've got Einstein's theory—and no doubt in the fullness of time we will have a better theory still. And once a scientist starts saying this is the ultimate theory, this is how the world is, we won't hear anything against it, of course it ceases to be science. And I think the strength of science is that it should remain forever open in that way.

That doesn't stop individual scientists of course being overwhelmed with hubris. However, I've got some sympathy with the likes of Richard Dawkins because they have such a struggle with the creationists that unless they state their case really forcefully, as the other side does, it looks like they're wriggling around when in fact their arguments are fairly secure.

You mean the scientific arguments are fairly secure but that some of the philosophical conclusions that they have drawn from them are not in fact scientific theories at all and are open to challenge by anyone—myself for example—a lay person in these matters ...

That's right, and I often do challenge these people. I once said that science can only deal in the facts and religion can deal in the interpretation of those facts. Dawkins and I might take the same lot of facts and draw very different conclusions and we should be allowed to do that, but I don't think Dawkins or I should present our interpretations of scientific facts or theories as anything like the truth. We shouldn't present them in the same manner as the scientific facts themselves.

In 1996 a meteorite from Antarctica was thought to contain evidence of life on Mars. It now looks as if it can be discounted. Are you disappointed about that?

Not really, because I think if we find life on Mars there's a good chance it's just cross-contamination from Earth. This is very frustrating because I see the existence of extra-terrestrial life arising independently of life on Earth as the absolute acid test for my whole view of a purposeful universe. Such a discovery would provide enormous circumstantial evidence for the purposefulness of the universe, and that the general trend from simple to

A purposeful universe

complex with the emergence of life and consciousness is written into the universe in a fundamental way.

The historian Thomas Reeves has written of your Templeton Address: "This address presents us with a classic example of a scientist whose yearning for God is blocked by the discipline that rules his mind and soul." Would you like to comment on that?
I think the two great strengths of science are its open mindedness and its unswerving intellectual rigour—the honesty that is brought to bear. I said at the end of *The Mind of God* that there are really two ways in which we people have traditionally sought truth. One is through rational reasoning and observation of the natural world, and the other is through some sort of spiritual or revelatory insight or mystical experience. I think that what I have to say refers only to the former. I believe that when we are dealing with the objective universe then we are quite right to employ those standards of rigour. To let the waters be muddied by what might be described as mystical yearnings I think would be quite wrong. That doesn't of course prevent people from looking inside themselves and seeking some sort of revelation.

Ironically I think that your books have encouraged the sort of new-age movement that wants a vague spirituality and belief in a cosmic power but doesn't want the demands and moral imperatives of a revealed religion.
Well you are absolutely right. Those sorts of people get joy from what I write. I can't help that of course. I write what I think is worth saying and all sorts of people plunder it for whatever they want. I myself recoil from this whole new-agey, wishy-washy mystical stuff—I've got no time for it at all. I try to distance myself from those people because I find the most fruitful exchanges on the science–religion front are not in that area at all, but in exploration of the question: Does quantum physics provide a way to God acting in the universe?

53. A universe from nothing? What could it mean?

By Stephen Ames

Our universe has come from nothing, according to renowned cosmologist Lawrence Krauss. This chapter considers his argument and says the universe is intelligible because it has been created by God.

At the Global Atheism Conference in Melbourne in 2012 I heard cosmologist Lawrence Krauss give a stunning talk based on his book *A Universe from Nothing: Why There is Something Rather Than Nothing*, (Free Press, 2012). His conclusion is worth quoting. He said, "I've told you three things. Firstly, you can get a universe from nothing. Secondly it will end badly. Thirdly, your life is far less important than you think it is." Krauss paused and then said, "So make the most of everyday you have in the sun." The auditorium of several thousand people went wild. They stood up, shouted, screamed, and cheered, all with sustained applause. An amazing moment.

Krauss tells a powerful and engaging story, of the last hundred years in astronomy, particle physics and cosmology. I won't attempt a summary of this story because I want to focus on two things. One is his "suggestion" as to how to get a universe from nothing. The other is his running commentary on theologians and religious people.

A universe from nothing—how? The answer is mind boggling—compared to our common sense view of things. I invite you to jump into the deep end and allow yourself to be carried along and see where it leads. Krauss assures us that a closed universe has zero total energy. The positive energy due to matter and negative energy of gravity exactly add to zero energy.

Secondly, physicists are working on combining the two outstandingly successful theories of general relativity (dealing with the very large scales of the universe) and quantum mechanics (dealing with the very small scales). It will mean that the rules of quantum mechanics will apply to space. One of the rules is that matter can "pop" out of empty space provided it disappears in a very short time so that it cannot be measured. This "popping" can be indirectly observed through the effects it produces. Krauss "suggests" that we can think of a tiny closed universe "popping" out of nothing, where here "nothing" means no space, no time, no matter, no energy. It then undergoes

A universe from nothing? What could it mean?

"inflation" leading to a big bang and the expansion we can observe, an expansion that started to accelerate about 10 billion years after the big bang. Krauss comments, "Does this prove that our universe arose from nothing? Of course not. But it does take us one rather large step to the plausibility of such a scenario." The plausibility draws on the claim that the proposal is consistent with everything we presently know about the universe.

One question Krauss addresses comes from the fact that the "nothing" he considers is governed by the rules of quantum mechanics. Where do these come from? His answer is that theoretical physics in various ways requires us to think of a more abstract "landscape" of a multiverse, in which universes are popping into existence with randomly different details of fundamental parameters and laws, never interacting. Some of these are "just right" to produce life, as is our universe—but this "fine tuning" doesn't need a Fine Tuner.

This looks like it should create a problem to do with Krauss' claim that we get something from nothing. He does not explicitly tell us whether this "multiverse" comes from "nothing." Were he to say the rules of quantum mechanics also apply to that "nothing" he will still be asked where the rules come from.

I think Krauss is basically asserting there is an ultimate brute fact that is fundamental—the "multiverse"—and everything flows from that starting point as he proposes. He need not get into the question of why there is something, not nothing. I think this is clear when he says that in our scientific understanding of the cosmos "the very distinction between something and nothing has begun to disappear, where transitions between the two in different contexts are not only common but required." The point is that even his most sophisticated version of "nothing" presupposes a "something."

We all heard about the accelerated expansion of the universe, when Professor Brian Schmidt at the Australian National University was recently awarded the Nobel Prize for its discovery in 1998. Krauss explains the universe will end "badly." Far, far into the future, the expansion will produce a dark cold sea of energy "with nothing in it to appreciate its vast mystery." This is a universe without meaning or purpose and for Krauss this is invigorating, "motivating us to make the most of our brief existence in the sun, simply because we are here, blessed with consciousness and with the opportunity to do so."

You may wonder whether Christians should have any interest in the work of scientists like Krauss. I think we should because I believe the universe is created by the living God as John 1, Hebrews 1, Ephesians and Colossians especially testify. I also believe that scientists are giving us deeper insight into God's handiwork, using their God-given powers. However, I wouldn't stitch the faith up to a particular stage of scientific inquiry, because inquiry moves on. There is a continuing need on our part to see the connections, to find the two-way resonances and dissonances between our continuing reception of the gospel and the daily exercise of reason, in science and elsewhere. It is also essential for our mission to communicate the gospel to people deeply immersed in science and technology, to show how being a Christian does not mean them giving up what God has given us in creation and in the power of human inquiry, fallible as it is.

So how might that play out with Krauss? Krauss has a lot to say about belief in God. He believes it is a form of wish fulfilment of what believers would like reality to be. Krauss thinks we ought to face reality. I can only agree with that imperative, but it is important not to be one-eyed about "reality."

Towards the end of *A Universe from Nothing* Krauss acknowledges that the idea of God as the "cause of all causes" means logically that you cannot ask "why God?" or "what created God?" His problem, however, is that there is no other evidence to support this idea of God. In my lectures I point students to several phenomena from the sciences for such support.

One concerns the phenomenon of human inquiry which we see worked out in one way through the natural sciences. Inquiry presupposes that what is inquired into is intelligible and open to rational explanation, but without prejudice to the forms of intelligibility or rationality that may be needed to fully understand. This is what gets inquiry going and keeps it going, as if it will continue. For no one thinks human inquiry will come to an end. This is the "horizon" presupposed in all human inquiry. *If* this is correct then inquiry would never grind to a halt in the face of a fact that explains all else but is itself without explanation—a "brute fact." Rather, the presupposition of inquiry could only be fulfilled were there something that explains the existence of everything else, whose very nature explains its existence. This is the idea of God, the "full horizon" intimated in all human inquiry. Then the universe gains its intelligibility from being created by God. Then human inquiry gains its presupposition from God, its imperative and its desire to know the truth from God. Then we could say to Krauss that God is much closer to him than he has so far recognised in what he loves so passionately.

A universe from nothing? What could it mean?

We could also say that how things will end, depends on this God. Of course, there is more to say in support of that "*if,*" but that is for another time.

Recently I was asked to speak after dinner to students at Queen's College at the University of Melbourne on "God and the Natural Sciences." To everyone's surprise 90 secular students attended. It was a great engagement for an hour. In one answer I spoke about Galileo's idea that God is the author of two books, the book of nature and the book of Scripture. If the two are correctly interpreted they cannot be in contradiction since they have one author. A student asked me what the God of the two books looked like. I offered several answers from the Bible but he kept looking at me as if seeking something more. So I spoke about his university studies and the expanding horizon of understanding exciting him and drawing him on. He nodded. So I said that there within his experience is an "icon" or window onto God, who is the author of the two books. He went "internal" for a moment and then nodded. It seems he felt he was well met.

54. Faith, hope and quarks: The search for God

By David Wilkinson

Is cosmology—the science of the origin and development of the universe—about an absence of God, or is it about a surer path to God?

We live in a culture where a conflict model between science and religion is very strong. However, within cosmology, my own subject, what we've seen is a number of folk in the public arena beginning to say there are questions which may go beyond science, which may go into philosophy and theology. In 1983 Paul Davies, distinguished populariser of cosmology, wrote the book *God and the New Physics*. Partly tongue-in-cheek he wrote, "In my opinion science offers a surer path to God than religion." Is cosmology about an absence of God, or is it about a surer path to God? That's what I want to explore.

Let's start with the universe. UDFj-39546284 is a fairly unremarkable galaxy in some ways; it's about one thousand times smaller than our own Milky Way. But what's really interesting about it is that this is one of the furthest objects we've ever seen in the universe. The light from the sun takes about eight minutes to make its journey to the earth; the light from this galaxy has taken 13.2 billion years. That means that as we look out into the universe we look back in time, because we're seeing the light that set off 13.2 billion years ago.

If the universe is awe-inspiring, then our current model for its beginning is just as awe-inspiring. The big bang model says that everything—100 billion stars in each of 100 billion galaxies, the whole lot—at some point was small enough to fit through the eye of a needle.

We think we can trace the history of the universe with reasonable confidence back to a time of 10^{-43} of a second old. Now, if you're not a mathematician that's a shorthand way of writing 1/(10 with 42 zeros after it) of a second. That's a very small fraction. If you're a biologist or an engineer or a normal person, you'll say, "Basically, that's zero, isn't it?" At that point, our current laws of physics break down.

You may be thinking, "Well, that's all well and good, but how do you know that's true?" Now, don't get me wrong; I'm not saying that the big bang is proved. In fact, in this kind of big science you rarely deal with proof.

Faith, hope and quarks: The search for God

What you have is an amount of evidence and then the best model which explains that evidence. The big bang model is pretty good, but it does have some problems to it.

Most embarrassingly for cosmologists is that we don't know what most of the universe is made of. About 23 percent of the universe, we think, is in the form of dark matter. We know it's there, because it exerts a gravitational influence, but we're not too sure what it is. And then 73 percent is made up of dark energy, and it's called *dark* energy because we don't know what *it* is. That means that we actually only know what about 4 percent of the universe is made of.

But there's one other problem with the big bang—a major problem. The problem is the limits of the two great theories of twentieth-century physics: quantum theory and general relativity.

General relativity, discovered by Albert Einstein, deals with things on the very largest of scales. It describes galaxies and the universe itself and does so, wherever you apply it, beautifully. Quantum theory deals with things on the very smallest of scales. It describes protons and electrons, and everywhere you apply it, it works beautifully.

Most of the time these two theories work independently, rather like different ends of the corridor in the physics department; they never ever meet, apart from at one point in the universe's history, and that's at 10^{-43} of a second.

A number of years ago Robert Jastrow, in his book, *God and the Astronomers*, talked about the frustration of our current laws of physics breaking down. He wrote:

> For the scientist who has lived by his faith in the power of reason, the story ends like a bad dream. He has scaled the mountains of ignorance; he is about to conquer the highest peak; as he pulls himself over the final rock, he is greeted by a band of theologians who have been sitting there for centuries.

Let me go on now and take Jastrow's cue: what are the big theological questions that may come from this?

The question of origins

Does the big bang prove God? You may have heard the type of argument that goes: If the universe began with a big explosion, the big bang, then who lit the blue touch paper? It must be God. Or, if the universe expanded from a little bit, then who put the little bit there first of all? God! In fact, it means that we replace the question mark in this type of argument by putting God in as the first cause of the universe.

Is this a good argument for the existence of God? I'm not convinced, for a number of reasons. The first is that this kind of cosmological argument has had a number of problems with the logic of it over the years. Can you use cause and effect and apply it to the whole universe itself? And what about the nature of time and asking what came before the creation of time?

But more importantly for me as a Christian theologian is that this kind of argument leads to difficult pictures about God. The first is what Charles Coulson called "a God of the gaps." Coulson said, "Beware if science has a gap in it, of inserting God into the gap, because as science progressively explains more and more of its own area, so God is pushed out into irrelevance."

The second is a problem of deism. A deist believes in a god who started the universe and then went off for a cup of tea, not to have anything more to do with it. But the images that the New Testament uses of God as creator are not of a god who simply started the universe off, but a God who holds the universe in the palm of his hand, keeping it in existence moment by moment.

Even with the possibility of a scientific theory for the very first moment of time, we are left with a question that goes beyond a god of the gaps type question. You see, if the universe arises from a quantum fluctuation leading to a rapid expansion, we're still left with the question: where does quantum theory itself come from? Where do the laws of physics themselves come from? And for me those are a reflection of this faithful God who sustains every moment of the universe's history.

The question of purpose

Does the big bang disprove God? You've probably heard this kind of argument as well: "The Bible says that the universe is the sovereign act of a creator God, physics says that the universe arose through a quantum fluctuation leading to a big bang." And people often say, "Now, which one are you going to choose?"

Let me use a silly illustration. What is a kiss? Well, a kiss is the approach of two pairs of lips, a reciprocal transmission of carbon dioxide and microbes and the juxtaposition of two orbicular muscles in a stage of contraction. That's a kiss in scientific terms. But when I get home and I see my wife, if I say to her, "Alison, let me get together with you for a mutual transmission of carbon dioxide and microbes. Let me juxtapose my orbicular muscle in a state of contraction with yours," she would say, "Get lost."

You see, with my wife I talk not about the carbon dioxide, I talk about meaning, value and purpose. Which is true? The description, in terms of science, or the description in terms of meaning, value and purpose? Both are true, but different, and to fully understand a kiss I need both. Therefore, for myself as a cosmologist and a theologian I'm quite happy to live with these two different descriptions of the universe.

Does the big bang disprove God? No, of course it doesn't, but we're left with the question of why, or the old philosophical question: why is there something rather than nothing?

The question of design

This question has been around for a long time in Western culture: "Is there anything that you can look at in the universe and find a proof for God as designer?" For the last few decades cosmologists have been playing around with questions of design—not proofs, but pointers. Why? Well, something called anthropic balances and the "Goldilocks enigma"; things seem to be just right in the laws and circumstance of the universe to make life possible.

Imagine we had a universe-making machine—it'd be very easy to construct. Two dials on it, one would set the expansion rate of the universe, the big bang force, one would set the force of gravity. In order to get a galaxy of structure like ours, these two need to be set to one part in 10^{60}. That's almost like if you're given a bow and arrow, blindfolded, spun round and asked to hit a target of one square centimetre on the other side of the universe. Don't get me wrong, I'm not then going to say, "Therefore, God." But for a number of colleagues, they look at these extraordinary coincidences and say there must be a deeper story to the universe.

The question of revelation

Finally: Why do I believe in a creator God? I don't believe in a creator God because of the pointers from the universe, although I find those fascinating. I believe in a creator God because I've encountered this God in Jesus of Nazareth, both in history and in experience. A God who self-reveals by becoming a human being. And this Jesus is the integrating point for believing that there's a creator and for holding together cosmology and creation.

Design, the Universe, Fine Tuning, Reality and Metaphysics

This chapter is an edited version of the Annual ISCAST lecture, given at Glen Waverley Anglican Church, Melbourne, Australia, in July 2013.

55. Fine tuning: Compelling evidence for God?

By John Pilbrow

The conditions for life in the universe are so finely tuned they strongly support a Christian understanding of reality.

We live in a remarkable 13.7 billion year old universe, on a 4.55 billion year old planet! Surprising as it may seem, it took some 10 billion years of cosmic evolution and the emergence of a vast universe for carbon based life even to be possible. There is strong evidence that special conditions, or fine tuning, have operated to enable a universe such as ours containing intelligent life to exist. This is sometimes referred to as the anthropic principle.

Britain's Astronomer Royal, Martin Rees, several years ago published a popular monograph, *Just Six Numbers: The Deep Forces that Shape the Universe* (Basic Books, 2009). Below is a brief (and, unavoidably, somewhat technical) summary of Rees's six numbers. No combination of any two or more of these six numbers has been found that can predict the value of any of the others.

Summary of Rees's six numbers

1. The ratio of the electromagnetic force to the force of gravity. When expressed as the ratio of the electrical repulsion between two protons (the positively charged particles in atomic nuclei) divided by the gravitational attractive force between them, the ratio is approximately one part in one followed by 36 zeros. If this number had been slightly smaller, the universe would have been short-lived and creatures would not have been able to grow larger than insects. Nor would there have been time for biological evolution to operate.
2. When two deuterons (heavy hydrogen nuclei) undergo nuclear fusion in the core of a star such as the sun to make a helium nucleus, only 0.7 percent (or 0.007) of the total mass is converted into heat that powers the sun. This number, 0.007, is an indicator of how firmly atomic nuclei bind together and it sets a limit on how long stars can live. Had the number been say 0.006 or 0.008 we could not exist.
3. The ratio of the average density of matter in the universe to the critical density recognised in cosmology. If this number were greater than 1, the universe would have collapsed a long time ago. If it were

very small, no galaxies could have formed. It means that the initial expansion of the universe must have been very finely tuned.

4. An anti-gravity effect to account for the observed accelerating expansion of the universe, first measured in 1998. Commenting on this, theologian Alister McGrath says "Fortunately for us (and very surprisingly to theorists), this is very small. Otherwise its effect would have stopped galaxies and stars from forming, and cosmic evolution would have been stifled before it could even begin." (*A Fine-Tuned Universe: The Quest for God in Science and Theology*, p. 28).

5. The ratio of the gravitational binding force to rest-mass energy ($E=mc^2$, where m is mass and c the speed of light). This is approximately 1/100,000 and is reflected in the tiny density variations observed in the background microwave radiation from the early universe. Had this number been any smaller, the universe would have been lifeless and not very interesting. On the other hand, a larger number would have led to a universe dominated by black holes where nothing else could survive.

6. Time is a dimension of space. It is necessary to think in terms of four-dimensional spacetime, consisting of the three spatial dimensions and time as the fourth dimension.

Alister McGrath's book, *A Fine-Tuned Universe*, referred to in the fourth of Rees's six points expands on his 2009 Gifford Lectures (http://www.abdn.ac.uk/gifford/lecture-texts/). He provides a theological reflection on the six numbers elaborated by Rees from within a Christian Trinitarian framework, and extends anthropic arguments to include chemistry, biology and the neurosciences. He argues that a theology of nature (or natural theology) should really be seen as a branch of theology. It is not about looking for evidence that God exists, but is about our exploration of nature within a Christian understanding of reality. Regarding St Paul's Areopagus address in Acts 17, McGrath comments that "Christian theology provides an interpretive framework by which nature may be 'seen' in a way that connects with the transcendent" (p. 11).

In setting the stage for consideration of fine-tuning, McGrath comments "For the theist, unsurprisingly, these observations point to the inherent potentiality with which the creator has endowed creation ... anthropic phenomena fit easily and naturally into a theistic framework, especially in Trinitarian forms." He argues further, "God made the cosmos with no

Fine tuning: Compelling evidence for God?

constraining influences other than the divine will and nature" (p. 120). Or again,

> The observation of anthropic phenomena is ... situated within a long tradition of theological and metaphysical reflection ... the general phenomena of fine tuning is consonant with Christian belief in a creator God, ... [and] that the most appropriate outcome for natural theology is to demonstrate that observation of the natural world furnishes conceptual resonance with, not deductive proof of, the Christian vision of God. (p. 121)

But is the anthropic principle merely a truism? The weak form of the anthropic principle, put very simply, is that we are here because we are here! The strong anthropic principle is more robust. It is the idea that the constituents of the material universe that began to emerge in the early stages after the big bang have an in-built order. This has led to greater levels of structure and organisation as newer and more complex features of the universe have emerged from elementary particles, through atoms and molecules to stars, planets, galaxies and the stuff of life. The natural world is thus held to possess law-like behaviour and in science we speak in terms of laws of nature. These codify observed regularities, but don't of themselves cause things to happen! While the laws of physics are well established, laws operating at higher levels of complexity in chemistry, biology or the neurosciences are not nearly so well understood.

McGrath correctly argues that chemistry, for example, cannot be understood solely from the properties of the constituent atoms. The fact that water is liquid from 0 to 100 degrees Celsius cannot be fully predicted from the properties of the single oxygen atom and the two hydrogen atoms that constitute a water molecule. It depends in part on the particular angle between the two hydrogen–oxygen bonds and the weak bonds between the two hydrogen atoms.

In outlining the nature of fine tuning in biology in some detail, considering the role of DNA and RNA and the remarkable role played by photosynthesis, McGrath turns to the work of Professor Simon Conway Morris from Cambridge University who has provided much evidence from the fossil record that evolution tends to find the same solutions time and time again. For example, he explains that the camera eye has evolved independently about six times while the compound insect eye as many as 50 times. While random mutations also play a role in evolutionary changes, what emerges is an overall directionality and it can be said that evolution "converges on a relatively small set of possible solutions for the problems and opportunities that the environment offers to life" (p. 193).

This is consistent with anthropic arguments that appropriate constraints have operated in all aspects of the evolution of the universe.

While some physicists and cosmologists postulate the idea of multiple universes, McGrath reminds us that the only universe we know anything about is the one we inhabit. Nevertheless he sees no problem for Christian theology in the multiverse hypothesis.

I can think of no better reason why we should seek to understand the special nature of our fine-tuned universe than the familiar words from John's Gospel, "And the word became flesh and dwelt among us (John 1:14)." Our God, whom we worship, became embodied in the very material of this world!

56. Does the "God Particle" bring us closer to God?

By Mark Vernon

In 2012 scientists discovered the so called "God particle," the Higgs boson. Mark Vernon considers whether reaching into the fabric of the universe is taking us closer to understanding the divine.

It began with a walk in the Cairngorms of Scotland 50 years ago. The physicist Peter Higgs had an idea about the origins of mass in the universe. The Higgs boson was born in the human mind. And now, after spending billions of dollars—as well as all the creative energy that cash represents—scientists at CERN (otherwise known as the European Organisation for Nuclear Research) have discovered the subatomic particle. Or at least, they have seen the signature for something close to what they expect the Higgs to be.

It has been called the "God Particle" because the Higgs plugs a crucial gap in what physicists refer to as the "standard model," which has been successful at describing the behaviour of matter, energy and forces. Yet the discovery of the Higgs does not mean that physics is now over. Far from it. In the 50 years since Peter Higgs' brainwave, cosmologists have discovered that the majority of the universe is made of stuff most probably unknown to science, the so-called dark mass and dark energy. The standard model will not be standard science for future generations.

Given those qualifications, it is striking that the Higgs has generated so much hype. This is partly because CERN, the organisation responsible for the Large Hadron Collider tests which discovered the particle, needs to justify the spending. So teams of spin doctors ensured the experiment produced regular headlines. But that only raises a further question: why are we so gripped by the weirdness of the subatomic world?

Physics powerfully resonates with the notion of cosmic design. Physicists look for theories that can be described using mathematics. When tested, these theories reveal the hidden nature of reality. The mathematical and hidden element readily fires the theological imagination. As the philosopher Gottfried Leibniz put it: "When God calculates and thinks things through, the world is made."

It is as if science and religion are part of the same enterprise: revealing the ways of God. "Science appears as a collective effort of the Human Mind to reach the Mind of God," writes the physicist and priest Michael Heller. "The Mind of Man and the Mind of God are strangely interwoven."

It is a powerful intuition explored in a famous essay by the Nobel Laureate for physics, Eugene Wigner. His title says it all: "The Unreasonable Effectiveness of Mathematics in the Natural Sciences." Wigner describes the descriptive and predictive power of mathematics as a "miracle." He continues: "It is hard to believe that our reasoning power was brought, by Darwin's process of natural selection, to the perfection which it seems to possess."

It is not so hard to believe if you believe that human beings are made in the image of God. The trumpeting of the discovery of the God particle flirts with the thrilling thought that we have taken one step closer to divinity. As the essayist Annie Dillard has written: "What is the difference between a cathedral and a physics lab? Are they not both saying: "Hello?'"

But it is easy for the theological imagination to become overexcited by science. For one thing, it seems likely that life can only emerge in a universe in which matter and energy are patterned and constrained. It is this patterning that gives mathematics a grip on nature. A life-bearing cosmos would inevitably be a mathematics-friendly cosmos.

Alternatively, you can ask what kind of God is revealed by the "unreasonable effectiveness" of mathematics. William Blake reflected on the deistic divinity implied by such an understanding of the cosmos and found it "soul-shuddering." His "dark satanic mills" are the impersonal cold machines that "grind out material reality," much as the indifferent Higgs boson is said to generate mass. A tyrannical God would fix life according to laws of nature, Blake continued. A world of such determined domains of space and time would be a prison. Further, the scientist or theologian who overplays humankind's capacity to understand the cosmos risks idolatry. "He who sees the ratio only," Blake mocked, "sees himself only."

More generally, you could say that contemporary physics so captures our imagination because it is the way many now do metaphysics. As the scholastic theologians of the medieval period gazed towards heaven seeking the divine, so we gaze into the heavens seeking replies to our great questions. Both activities promise to reveal the nature of reality and so something of our own nature.

Only perhaps we need to learn not to be so concrete, so literal. If it is a mechanism you seek, then the God particle will thrill you. If it is life, then clues must be sought somewhere else.

Does the "God Particle" bring us closer to God?

This chapter is a slightly edited version of an article that first appeared in the 7 July 2012 edition of the international Catholic weekly, *The Tablet*, and has been reprinted with permission of the publisher. See www.thetablet.co.uk

57. God a "combination of love and mathematics"

By Madeleine Davies

A "cultural Anglican" met a practising one in an Oxford forum in February 2012. Richard Dawkins and the then Archbishop of Canterbury, Rowan Williams, revealed how much they had in common.

The University of Oxford's theology faculty organised their public dialogue in the Sheldonian Theatre. Dr Dawkins, until 2008 Oxford University's Professor of the Public Understanding of Science, confessed to having sung a hymn in the shower that morning, and described himself as a "cultural Anglican," even an agnostic, after admitting that he couldn't be 100 percent sure that God did not exist.

Dr Williams confirmed his belief in evolution. He said that he had been "inspired" by the professor's books: a "delight to read."

The 90-minute dialogue on "The Nature of Human Beings and the Question of Their Ultimate Origin" was chaired by the philosopher Sir Anthony Kenny, a former Roman Catholic priest, now an agnostic.

Dr Williams and Professor Dawkins discussed the possibility of multiple universes, whether consciousness was an illusion, and why tragedies occurred.

Much of the debate focused on consciousness, which, Dr Williams suggested, had so far defied scientific explanation. Although he agreed with Professor Dawkins that the evolution of humans was gradual, he defined the beginning of humanity in the image of God as the moment when a human became conscious. This included consciousness of the divine.

Professor Dawkins agreed that consciousness was "deeply mysterious," but suggested that it would be solved by a combination of neuroscience and computer science. He acknowledged that there might have been instances of "sudden process" in evolution, such as the acquisition of language and the emergence of DNA.

The two also discussed free will, an idea that, some scientists believe, has been undermined by experiments in neuroscience which indicate that decisions are taken by a person's brain before he or she is conscious of them.

Later in the debate, an audience member asked whether the fact that human beings were "immensely imperfect" was evidence of the failure of evolution or design. She gave the example of children who die before they could fulfil their potential.

"It's tough: stuff happens," Professor Dawkins replied. "Death before reproduction is what natural selection is all about and it's tragic. If one looks around the world and sees the sheer amount of suffering, it is exactly as you would expect it to be if it were just blind forces of nature happening, if there were no overarching process."

Dr Williams said that he was "obliged to say a little more." While declining to offer a "mega-theory," he said that "in a world where change and chance are uncontrollable, tragic accidents happen." He confessed that this was an "elegant version" of Professor Dawkins' blunt response.

The two also discussed the biblical creation story. Professor Dawkins confessed: "I am baffled by the way sophisticated theologians who know Adam and Eve never existed still talk about it as if it had some profound wisdom for us in an allegorical sense." He suggested that theologians "waste their time" by reinterpreting it for the twenty-first century.

Dr Williams argues that he could turn to Genesis to "understand my moral and spiritual place within the universe."

Latterly, Professor Dawkins confessed that he could not be "absolutely confident" that God did not exist, and described himself as a "6.9" on a one-to-seven scale of belief, because the probability of God was "very, very low."

He described Darwin's discoveries as an "astonishing thing to have happened." "If you say 'I expect there was a God anyway,' that completely undermines the whole rationale for doing it," he argued. "It's a betrayal of everything that science stands for."

For Dr Williams, God was "what sustains the entire system." For Professor Dawkins, he was something "messy" that people used to "clutter up their world-view."

This, for Dr Williams was the one thing that "seriously divides" him from the professor. "God is not an extra to be shoehorned into it," he said. "The elegance and beauty [of the world as revealed by science] is simply framed by a sense of a God who is a combination of love and mathematics."

This is a slightly edited version of an article which appeared in *Church Times* on 2 March 2012. The debate can be seen on YouTube.

58. Gravitational waves discovery opens new way of looking into the universe

By Stephen Ames and John Pilbrow

The recent discovery of gravitational waves is an exciting moment for science, but what does it mean for theology?

The first direct observation of gravitational waves, reported officially in the world press on 11 February 2016, has provided confirmation of a very important prediction made by Einstein in 1916 as a critical test of his theory of general relativity.

Our everyday lives take place in the Newtonian world involving time and three space dimensions, and where gravity is an attractive force between objects. Newtonian physics is good enough to put people on the moon, and to cope with the everyday world. However this was turned on its head in 1905 when Einstein introduced special relativity and the seemingly bizarre concept of four-dimensional space-time where time depends on the motion of the observer. While special relativity deals with objects moving at close to the speed of light, something more was needed to be able to understand what happens near massive bodies. Einstein published his theory of general relativity in 1915 where gravity is pictured as the curvature of space-time. It reveals a way of looking at the world very different from our everyday experience. Readers may be surprised to learn that corrections due to general relativity are built into the operation of GPS satellites!

In 1916 Einstein was able to show that general relativity could account precisely for the observed anomaly in the perihelion of Mercury (i.e., the point of closest approach to the sun) that Newtonian physics could not explain. Notwithstanding that success, he knew that general relativity needed further testing and made the following three predictions: (1) Light from distant stars would be deflected as it passed the sun (confirmed in 1919); (2) Light passing a massive star would be shifted towards the red end of the spectrum (actually confirmed in 1959 using gamma rays in an earthbound experiment); and (3) gravitational waves, or ripples in the fabric of space-time, which should travel at the speed of light, caused by the motion of very massive bodies (recently announced).

Gravitational waves discovery opens new way of looking into the universe

The science behind the detection

Gravity is by far the weakest of the four fundamental forces in nature, the other three being electromagnetism, and the strong and weak nuclear forces. This means that extremely sensitive apparatus is needed to detect the tiny changes in gravity caused by some of the largest high energy events in the universe.

The extreme sensitivity needed to detect these waves is provided at two identical LIGO facilities (Laser Interferometer Gravitational-Wave Observatory) in the USA, one in Washington State and the other in Louisiana, just over 3000 km apart. At each site a pair of four-kilometre-long laser beam "arms" (in a vacuum) are placed at right angles; each arm has two specially coated mirrors at the ends. The mirror coatings were produced by Australia's CSIRO.

The length of the arms is changed by gravitational waves which shorten one arm and lengthen the other. The measured difference is 10^{-18} metre or 1/1000th the size of a proton. That level of precision in human measurement is itself amazing. In September 2015, a single event was detected, first at one observatory and then 6.9 milliseconds later at the other with exactly the same time profile.

After five months of analysis of the data, the results were published on the day of the announcement, Thursday 11 February 2016, in *Physical Review Letters*. The article is very readable, and begins with an informative short history of the search for gravitational waves. Its main focus, however, was on the construction, operation and testing of the LIGO facilities at both locations and outlining the steps in data analysis that established there could be only one explanation for the observation of the event 1.3 billion light years away. This was the spiralling in of a pair of black holes, one 36 times the mass of the sun and another 29 times the mass of the sun, coalescing into a single rotating black hole.

It is more than 50 years since retired New Zealand mathematics professor Roy Kerr discovered an exact solution of the equations of general relativity that predicted rotating black holes. For this work he was named as one of three recipients of this year's Crafoord Prize, a prize awarded in fields not covered by Nobel Prizes.

Detecting gravitational waves is a truly stunning scientific achievement, theoretically but especially empirically, with the design, construction and operation of the amazing LIGO observatories. All this drew together a vast amount of scientific, technological and engineering capabilities and considerable international cooperation.

Prediction in science

On the basis of Einstein's theory, researchers knew the kind of signal they were looking for. This is reassuring that the science is on the right track. Physicist and Anglican priest John Polkinghorne has observed that science results in "a tightening grip of a never completely comprehended reality." Of course he doesn't mean that reality is somehow confined to what science can tell us.

The discovery hasn't brought into being a new theory, but rather is a brilliant confirmation of Einstein's third prediction. Researchers will have even greater confidence in Einstein's general relativity now, but they haven't waited until now to use it. It has underpinned thinking in cosmology for a century and will continue to do so. As with all good theories, if they have been confirmed so far as testing has allowed, then that is a basis to believe them to be the best available to date, not necessarily the last word. It is wonderful when predictions are in fact fulfilled and with such exacting precision.

This success will spur on new research. This new gravitational wave astronomy will be looking for other examples, possibly rotating neutron stars. Again, physicists have a theoretical picture of what gravitational wave form they expect to find coming from the early universe, looking back further than 13 billion years.

The present result also highlights again the unfinished task of unifying general relativity and quantum mechanics which so far seem incompatible, yet both are involved in the theory of the big bang. String theory has been the strong program aimed at achieving this unification. It is also associated with the multiple universe proposal. Physicist Lee Smolin and philosopher Robert Ungar in their book *The Singular Universe and the Reality of Time* reject the multiverse idea as non-scientific since there can be no empirical testing of the proposal, a view strongly supported by Paul Davies and George Ellis. By way of contrast, the extraordinary empirical detection of gravitational waves, like the discovery of the Higgs particle in 2012, emphasises the centrality of the empirical testing of theories to the natural sciences. However it appears that many researchers working on string theory step back from the requirement of empirical testing.

Science, philosophy, theology

Whenever there is a major scientific breakthrough that causes headlines in the world press, there are Christians who jump on the discovery to say it provides additional reasons to believe in God or evokes doubts. Also, some atheists will use the same information to claim that it is another nail in the coffin of the Christian God.

Such responses miss an important point. Science is about exploring the world, with theories that do not make any reference to God. There is no direct link from scientific discoveries to their theological meaning. So, no need to jump to negative conclusions and no need to dismiss new scientific discoveries because they seem not to immediately fit our view of God. As J. B. Philips said long ago, maybe our (idea of) God is too small.

Gravitational wave astronomy opens up a new way of looking into the universe. It signals a new form of astronomy adding to what optical and radio telescopes have already allowed us to see. This naturally recalls 1509 when for the first time Galileo turned his telescope to view the planets of our solar system. There was a great ruckus with church authorities silencing Galileo. But the issue was not about what Galileo observed through his telescope. The Jesuit astronomers reproduced all his scientific observations. Pope Urban VIII accepted all the observations. The issue was what the observations meant. The observations certainly showed that the old earth-centred view of the world was incorrect. However, at that time, they did not show the earth moved around the sun and one reason for this was that all the observations were accommodated by a new earth-centred view in which the sun, orbited by all the other planets, itself orbited the earth. One of the other objections to the sun-centred view of the world was the universe would have to be unimaginably, unbearably vast to accommodate the fact that there was no difference between the angle of a star observed from the earth, say during Spring, and the angle observed six months later during Autumn, after the earth had travelled (supposedly) half way through its orbit of the sun.

Everyone, believer or not, is still coming to terms with the scale of the universe and the discovery of gravitational waves is the latest finding to bring home the scale and other properties of the universe. The vastness of the universe and the billions of galaxies each with billions of stars leads to wondering whether there are other civilisations on other planets circling stars. NASA is finding many, many planets that are contenders. Is it still the

Design, the Universe, Fine Tuning, Reality and Metaphysics

case that "new born worlds rise up and adore," as the hymn puts it, or are we alone? Both views are argued. The possibility of multiple worlds is not new to theology. It was part of the thinking of the fifteenth-century cardinal Nicholas of Cusa, theologian, philosopher and astronomer.

The observation of gravitational waves is a wonderful scientific discovery, the outcome of a great international cooperative endeavour. This is an example of "big science" at its best and stands in a long line of several large-scale international scientific endeavours that include the Human Genome Project, the work of the IPCC on climate change, and the discovery of the Higgs boson in 2012. These all testify to the great if still fallible power of human inquiry both theoretical and empirical. It is one sign of human beings at their best and shows the power of seeking understanding for its own sake, without losing sight of the possibility of other useful benefits.

The 1662 Prayer Book teaches that we can do nothing good without God. If so, are the processes of scientific inquiry, which make no reference to God, either in theory or in practice not good? Or is God somehow at work in a hidden way in scientific inquiry? We prefer to say that scientific inquiry is certainly fallible, but what it produces is a positive good, however it may subsequently be used. Therefore we prefer to reflect on the practice of scientific inquiry to discern there the hidden presence of God. Does scientific inquiry really intimate the presence of the transcendent God? Is this arguable or just a delusion?

As the above discussion showed, the discovery of gravitational waves was the result of both theoretical and empirical inquiries. The theory is a century old this year. Einstein's argument is still valid. The conclusion follows logically from the premises and unless someone can point out flaws in the premises or steps in the argument, the conclusion holds. It is virtually unconditioned by time, history, culture, wealth, personality, gender, and even by the standards of the science of a hundred years ago. Something that transcends all these conditions comes to light in Einstein's argument (and all such valid arguments from premises to conclusions). This is just one of the intimations in scientific inquiry of a transcendent reality, which is commonly designated "God." This line of thought is supported by works such as John Haught's *Is Nature Enough?*, Michael Polanyi's *Personal Knowledge* and chapter 3 of *Miracles* by C. S. Lewis.

If this seems a bit odd scientifically and religiously, consider Peter Harrison's *The Fall of Man and the Foundations of Science*, showing that the leading lights of the new approach to nature in the sixteenth and seventeenth centuries saw the emerging experimental method as correcting the damage to our senses and our reason due to the fall. Their aim was to recapture Adam's lost knowledge of nature!

Of course, there is more to say on all the above. Like our Christian forebears we look to the incarnate Logos through whom and for whom all things are created—including, it seems, gravitational waves—to open our eyes to what all this means.

59. How science can help us understand prayer

By John Pilbrow

Can quantum physics or chaos theory shed any light on what happens in prayer? These and other questions are considered in *When I Pray, What does God Do?*, the latest book by Methodist theologian David Wilkinson, Principal of St John's College, Durham University, UK.

Many readers will resonate with chapter 1 where Wilkinson addresses his own problems with prayer. He has tried to follow all the advice about prayer on offer, only to find much of it rather frustrating. He does not offer easy answers to difficult questions, nor does he attempt to provide a definitive answer to how God answers prayer. Rather this book "is more a record of a personal and an ongoing journey of how a Christian, who wants to take both the Bible and science seriously, begins to think about these things." With doctorates in both astrophysics and theology, Wilkinson is well-placed to explore insights from modern physics that may help us understand something of God's role in prayer.

The author deals honestly with life's realities, particularly in the face of apparently unanswered prayers, such as when a loved one is found to have a debilitating illness. His wife, also a Methodist minister, suffers from rheumatoid arthritis. She has not been healed though much prayer has been offered on her behalf. This he says "intensifies the difficult questions concerning prayer."

Wilkinson recognises that science and suffering both pose really difficult questions as to how God answers prayer, and that the Bible does not give simple answers. He goes on to say,

> My own personal experience of science and, to a lesser extent, suffering, makes me question a whole number of aspects of prayer and how God answers ... my experience of God in the Bible and in my daily life continues to encourage me to pray expectantly that God will act ... this I hope, is not because I simply live in two different worlds, shutting out the difficult questions by immersing myself in loud chorus and sheltering in a bubble of Christian subculture.

Chapter 2 is an exposé of seven everyday myths about prayer. These include the slot-machine-of-faith God, the health-and-wealth God and the are-you-ready-for-a-miracle God. Wilkinson resonates with the late Henri Nouwen's observation that until we dispose of the false gods of popular religion, we cannot truly respond to the God of the Bible.

In chapter 3, Wilkinson seeks to provide a context for praying in the contemporary world, consistent with the practice of Christian prayer down the ages, by invoking many familiar prayers from both Old and New Testaments. For example, in the book of Nehemiah, prayer is about the political situation. On the other hand, the Gospels emphasise praying for friends in need and praying with faith. As already noted, sometimes prayers return unanswered, and while the Bible's teaching here is not straightforward, it is made clear that "God has the power to do whatever he likes."

He goes on to say,

> As a scientist, I do wonder that if in our prayer we so rush towards the miraculous that we do not give enough attention to thankfulness for creation … As well as the natural seasons I would like to see more churches thank God for the gift of science, with the associated joy of exploring and shaping the world.

In the light of this he reflects on the plight of many scientists within the churches who sense that their work is a calling from God, but who do not feel that the church recognises or values this calling.

In the next chapter, "Out of Date Science and the Problems with Miracles," Wilkinson is rightly critical of much contemporary theology that remains locked in a Newtonian (or clockwork) universe, a highly predictable world in which biblical miracles are considered violations of the laws of nature. In opposing such views, Wilkinson observes that "the whole point about miracles in the Bible is that they are not about God's hidden action … quite the opposite, drawing attention directly to God's special actions in the world for specific purposes." He would like to see theologians in this century at least catch up with the science of the last century.

Within a discussion of the mind–brain problem Wilkinson highlights an important insight from Christian philosopher, Nancey Murphy. That is, "God relates to the whole human person rather than just a 'god receiver' that mysteriously resides within us."

With his astrophysics background in mind, the author turns to consider counter-intuitive insights from quantum mechanics and chaos theory from twentieth-century physics. The unpredictability of the quantum world applies to the very small, such as electrons and protons. If we know where they are, we cannot possibly know how fast they are moving, and vice versa.

This is expressed in the famous Heisenberg uncertainty principle. Then there is the even more counter-intuitive idea that particles that have once been in contact can still influence one another even when separated so far apart that a light signal cannot pass between them. Wilkinson considers whether this gives "freedom to God to work within the scientific laws of the universe in unusual ways."

Secondly, unpredictability also lies at the heart of modern chaos theory. This arose in the context of weather forecasting where computer modelling showed that small changes in just one property could produce large-scale effects elsewhere, sometimes called "the butterfly effect." Wilkinson thinks this could be more significant than quantum theory because it is dealing with systems at the everyday macro scale. So, when we pray, perhaps God responds by working through the openness of a chaotic system.

In the non-Newtonian world of quantum mechanics and chaos theory, a more nuanced view is required of the laws of physics since these, at face value, simply describe "the regularities that we have discovered about the universe." Wilkinson explains that "some phenomena appear miraculous not because they are breaking scientific laws but simply because they reflect a deeper, truer reality" that we don't yet understand. In this light it is considered that the universe itself has been endowed with a certain level of freedom. Therefore, he says, "if God has granted some freedom both to human beings and the physical process it would be difficult to understand if he did not reserve some freedom for his own action." We then must consider how God's freedom relates "to his sustaining the laws of physics, his sovereignty in drawing the creation to new creation, and in respecting the freedom that he has granted."

Though some things in this book may prove challenging, there is also much practical wisdom. Two examples will suffice. First, Wilkinson is concerned that much of our praying is about issues that seem a trivial waste of God's time. Should we, for example, pray for a parking space at the supermarket? He queries, "Why produce a parking space for a Christian who could probably benefit from a little walk by parking further away?" A second example concerns a man who sought prayer and laying-on of hands for healing. Afterwards, on being asked whether or not he had seen his doctor, the fellow replied, "That's not very spiritual" to which Wilkinson retorted, "to go to see a doctor is very spiritual! ... The skill of the doctor is made possible by being made in the image of God, and the human body's own powers of recovery once again are made possible by God."

How science can help us understand prayer

Wilkinson wonders if when we pray we tend to focus too much on what we are doing ourselves, often with far too many words. He says that given the large qualitative difference between us and God, when we pray our focus should be on God and not ourselves.

The final chapter is about praying in the light of what God does. While we can be sure that God sustains the laws of nature and responds to our prayers, his actions also include transforming this creation into new creation and transforming the person who prays to collaborate in building God's kingdom.

In conclusion, this is not a prayer manual but a book that provides good reasons to go on praying in today's scientific and technological world. The author's appeal to the unpredictability at the heart of both quantum mechanics and chaos theory provides pointers towards, but not proofs of, what God does when we pray.

Other matters addressed in the book but not touched on here include so-called natural evil, why we shouldn't make a distinction between the natural and the supernatural, and whether it is appropriate to end our prayers "In the name of the Father, the Son and the Holy Spirit."

The foreword by the Archbishop of York, John Sentamu, is a beautifully crafted endorsement of the book. He urges us to take its message to heart, and so do I. *When I Pray, What Does God Do?* is published by Monarch.

60. In the "great silence" can we be sure we are alone?

By Jonathan Clarke

If we are not alone in the universe, why have we not "heard" from other civilisations? Does humanity's future rely on increasing advances in technology?

Two of the great questions in space science and astronomy are "Is there life beyond Earth?" and "Are we alone?" The first question is the province of astrobiology, the science of life in the universe, of which so far the earth provides the only examples. The second is the subject of the Search for Extraterrestrial Intelligence, abbreviated to SETI. Both have been speculated about since the ancient Greeks and more recently have been the subject of scientific research. Both astrobiology and SETI raise fascinating philosophical and theological questions.

The peak of optimism about both SETI and astrobiology was perhaps in the eighteenth century, when most scientists thought that life was universal. Sir William Herschel (who discovered the planet Uranus) even thought there might be life on the Sun. But as our knowledge of the scale and complexity of the universe increased, the likelihood of life, at least in our solar system, has diminished. At present, astrobiology aims to find potentially present or past habitable environments on Mars, Europa, Enceladus and perhaps elsewhere. But "potentially habitable" has yet to translate into "definitely inhabited," despite some ambiguous clues from Mars.

Finding life elsewhere in our solar system, especially if proved to have an independent beginning, would have major implications for finding it elsewhere in the universe. If life occurs independently on two (or more) bodies in our solar system, or even once did, then life is almost certainly commonplace in the universe. If not then it may not be so common, though not necessarily rare.

Finding of life elsewhere in the universe raises the so-called Fermi Paradox. If life is common, the paradox goes, if intelligence is the inevitable product of life, and technology the inevitable consequence of intelligence, then where is everybody? Surely in a 13 billion year old universe of a hundred billion or more galaxies, each with a hundred billion stars, some species must have developed the means of travelling or signalling across

In the "great silence" can we be sure we are alone?

interstellar distances. But they haven't done so as far as we know, despite decades of efforts by scientists involved in SETI.

This so-called great silence has many explanations. Perhaps life is actually rare. Perhaps intelligence or technology are not inevitable. Perhaps there are limits to technology that prevent travel, communication, or even showing signs of presence across interstellar distances. Some postulate a great filter, or several, that greatly reduce the likelihood of a species emerging onto the interstellar stage. Life may require much more specialised conditions that we know of, be more subject to natural catastrophes than we realise, or perhaps technological societies nearly all self-destruct after all. This is bad news for trans-humanist philosophers whose hope for the future is a vision splendid of biological engineering, fusion of the biological and the machine, and uploading of human consciousness.

Such speculations have led the trans-humanist philosopher Nick Bostrum of Oxford to write: "I'm hoping that our space probes will discover dead rocks and lifeless sands on Mars, on Jupiter's moon Europa, and everywhere else our astronomers look. It would keep alive the hope for a great future for humanity." The more common life, especially advanced life, is in the universe, the greater the likelihood that the great filter lies in the future. Alternatively, as suggested by Graham Phillips, while planets are common in the universe (in 2011 we saw the discovery of the 500th extrasolar planet), maybe it is civilisations that are rare. The universe is certainly full of planets, life may be common, but perhaps it is the business of civilisation that is hard to start. Perhaps we can pin our hopes in this.

Such views are, in my view, premature. We have hardly begun to explore the other worlds of our solar system; our ability to do so is very limited and in many cases we don't yet really know the sort of questions we should be asking when looking for life potentially very different from our own. Sufficiently detailed scientific exploration to answer this question may take centuries of research, especially when we consider that the surface area of Mars is equivalent to all the Earth's continents combined. To show that the other worlds of our solar system have always been lifeless will take centuries more. Even the evidence for the earliest life on Earth is hotly disputed, evidence for or against past life on Mars even more so, and is far more difficult to investigate.

Likewise, while our telescopes can see back almost to the first moment of time, we are currently unable to detect a possible technological civilisation such as ours, over distances of more than a few light years. The "great silence" may be more a result of our inability to hear than telling us something about the universe. Furthermore, the many assumptions about astronomy, life, intelligence and technology that underlie the Fermi paradox may be deeply flawed. Until we know a lot more about the universe (and have gained a lot more experience in understanding and managing the consequences of a technological society) we simply do not know enough to draw any such conclusions.

However, such speculations do serve to highlight the assumptions and implications of particular metaphysics. If someone's eschatology is pinned on technological transcendence, then the chance that salvation does not lie there will be profoundly disturbing. Even if the dreams of the transhumanists can be achieved, they do not offer an escape from the inevitable death of the universe. Whether it is freeze or fry, this universe will eventually come to an end, finishing the hopes of all whose salvation lies within its confines, no matter how advanced its technological progress may become.

Our Christian hope is different. It is not built on science or technology, even though both grow out of being made in God's image. We recognise the finiteness of this world, even though its history may stretch over billions of years and stars may number in the trillions. Our hope is based on following Jesus, through whom all things will be brought together to their goal, not only in this world, but in the next. Not simply in a future "heaven," but in a whole new eternal creation, in which all things find their fulfilment, not just humanity. The Apostle Paul speaks of all creatures in heaven, on Earth, and under the Earth, brought together under Jesus as Lord. Expressed in the three-decker cosmology of the Bible, this is a promise of all beings in the universe, past present and future, recognising Christ's sovereignty. Whether this involves just humanity and the hosts of heaven, or an entire universe of creatures—perhaps even Klingons or Sontarans—it will be a sight to which we can look forward.

61. Quantum uncertainty and the action of God

By John Pilbrow

Quantum physics may give us a glimpse into understanding how God acts in the physical world.

Experimental science began to flourish some 400 to 500 years ago at a time when people rediscovered something of the spirit of what Jesus taught when he said "You will know the truth and the truth will make you free" (John 8:32). The early scientists considered science to be a worthwhile endeavour and, for some, even a Christian vocation. Such exploration of the natural world and discovery of the laws of nature makes sense only because of God's faithfulness in sustaining the creation. Similarly, we may discover something of the nature of God by faith, through the revelation of God in Jesus Christ. Science and faith (and theology) provide complementary ways of looking at the world and our experience of it.

Until the discoveries of relativity and quantum mechanics early last century, science appeared to support the idea of a "clockwork" universe where everything was determined by the laws of nature, a view that is still widely believed today outside science. However, the closed mechanical system of classical physics has given way to an understanding that God created the universe out of nothing in such a way as to be open to his ongoing action, a view that takes seriously the age of the universe and the evolutionary character of unfolding cosmic history. It is also recognised that emergent higher level properties exist as we go from physics, through chemistry and biology to the neurosciences. The "laws" that apply to ever more complex systems are far from being fully understood because the complex wholes are much more than the sum of their parts. For example, physics cannot be reduced to mathematics, biology cannot simply be reduced to physics and chemistry and, in particular, brain function is thought to involve much more than physical and chemical processes. Emergent levels of meaning and order in the hierarchical structure, however, present possibilities for gaining insights into both bottom-up and top-down action.

Quantum mechanics, which applies to the world of atoms and elementary particles, does not allow us to predict the outcome of particular events. Through the Heisenberg Uncertainty Principle, we can begin to understand why if we know how fast an electron is going, we don't know where it is.

Design, the Universe, Fine Tuning, Reality and Metaphysics

Quantum uncertainty also applies to radioactive decay. We cannot predict which particular radioactive atom will decay next, but we know how long it will take for half of them to decay (half-life). Radioactive dating methods using carbon 14 (half-life of 5730 years) work up to 60,000 years, whereas uranium to lead decays yield dates for ancient minerals from the earth's crust as high as 4.4 billion years.

Might quantum uncertainty provide a way to understand aspects of God's action in the world through bottom-up processes? If so, it would be necessary to explain how microscopic quantum effects, at the level of atoms or elementary particles, could bring about changes on larger scales. Would this be sufficient to provide scope for God to influence the course of events without disrupting the structures of nature, yet at the same time ensuring they will continue to provide for novelty and regularity in the world? To seek to understand how God might work through natural processes is a perfectly legitimate endeavour. Thus physicist-theologian Robert Russell, in considering whether quantum uncertainty is involved in biological evolution, asked: "To what extent do point mutations arise from the interaction of a single quantum of radiation and a single proton in a hydrogen bond in a specific base [in DNA]?" (*Evolutionary and Molecular Biology: Scientific Perspectives on Divine Action*).

From 1991 to 2003, six conferences on Scientific Perspectives on Divine Action took place, jointly sponsored by the Vatican Observatory and The Center for Theology and the Natural Sciences in Berkeley. They represent the most significant collaboration ever undertaken at the science–faith interface. Scientists, scientist-theologians, theologians and philosophers explored ways in which God's action in the world might be understood in terms of modern scientific paradigms, with emphasis on objective, non-interventionist divine action consistent with a universe that exhibits genuine openness. Quantum uncertainty as one possible basis for God's action was explored in some depth. On the other hand, the idea that God acts by intervening now and then was rejected as being inconsistent with the picture we have of God's constancy and reliability in the unfolding revelation of Scripture and in our Christian experience. Were God to act by suspending the laws of nature, God would be seen to be opposing himself, and there could be disastrous consequences for us. Top-down and bottom-up approaches to God's action were considered.

We can understand top-down effects through a familiar example such as moving our hand or foot. The intention starts in our minds. This leads to brain states that then interact with our central nervous system and, ultimately, lead to the movement of our hand or foot. This is a partial outworking of our God-given freedom constrained by the laws of the physical universe.

Throughout Scripture God is described as ever active, continuously upholding the universe, without violating the freedom given to the universe to be itself. This is referred to as general providence. God is also believed to act in specific ways through what is called special providence such as answers to prayer, healings, coincidences and particular events that we attribute to God by faith.

If quantum uncertainty provides the necessary linkage, it must influence upper levels of complexity through bottom-up quantum effects at the atomic or molecular level, while God's intentions are also being effected in a top-down manner through cosmological and evolutionary processes, on the one hand, and by direct interaction with us on the other.

On prayer, modern scientific understanding leads former Professor of Mathematical Physics at Cambridge, John Polkinghorne, to say,

> We can take with absolute seriousness all that science can tell us and still believe that there is room left over for our action in the world, and for God's action too ... Prayer is not magic. It is something much more personal, for it is an interaction between humanity and God.
> (*Quarks, Chaos and Christianity*)

In a deterministic universe, there would seem to be little room for effective prayer.

My hope is that most readers will find something helpful here. Cosmologist George Ellis says: "A very traditional Christian view can be vindicated which makes sense to the ordinary believer, and which can make sense also in relation to present-day science" (*Neuroscience and the Person: Scientific Perspectives on Divine Action*).

I have presented what is really a metaphysical argument in talking about the possible relevance of quantum uncertainty to God's action. The scientific details and theological implications have yet to be fully worked out. The universe is not the deterministic clockwork world as believed in the past, but exhibits its own God-given freedom. This means we too have a level of freedom as a gift from a loving God. This should affect our worship and make us more fully aware of the immensity of the created order.

62. Intelligent Design theory scientifically and theologically flawed

By Allan Day

Support for the Intelligent Design (ID) concept of creation is misguided.

Most scientists who are Christians would consider that science and faith are complementary approaches to truth and that the pursuit of science by appropriate scientific methods does not negate their understanding of God as creator and designer of the cosmos. They would understand science as exploring the secondary mechanisms used by God who is the ultimate cause. Thus, theistic evolution presents no problem for most Christians, Catholic and Protestant. Scientifically, evolution, understood as the progressive development of more complex biological life is strongly supported by the evidence. Indeed it forms much of the framework of modern biology.

Why then all the fuss about Intelligent Design? Is it simply a statement of the Christian worldview of God as creator, sustainer and designer? Well, no, it is not!

There is a body of Christians who consider evolution to be intrinsically materialistic and therefore any science which supports it is to be opposed as contrary to theistic belief. There has therefore been an attempt to discredit evolution on ideological grounds, not by the appropriate use of science, but on the basis that there is an alternative theistic science which can be pursued outside the basic scientific method. ID is the latest attempt to do this—to assert that creation science is science and should be pursued and taught in parallel with conventionally accepted science. It is perhaps not surprising that the ID concept has been opposed by both Christian and secular biologists. It is both bad science and bad theology.

ID is not simply the concept that there is a designer god behind the universe—a concept that, as a metaphysical worldview, is *consistent* with the findings of science, but neither *established* nor *negated* by them.

ID asserts that there are certain aspects of biology that exhibit "irreducible complexity" and therefore cannot be explained by evolutionary processes, nor indeed be subject to scientific investigation. The proper "scientific explanation" therefore is to insert design as a mechanism.

ID was first conceived by a meeting of anti-evolutionary philosophers, scientists and activists and promoted by William Dembski—a

Intelligent Design theory scientifically and theologically flawed

philosopher and mathematician—in his book *Intelligent Design:The Bridge Between Science and Theology* published in 1999; and by Michael Behe—a biochemist—in his book *Darwin's Black Box: The Biochemical Challenge to Evolution* published in 1996.

The opening words of Dembski's book provide his definition:

ID is three things, a scientific research program that investigates the effect of intelligent causes, an intellectual movement that challenges Darwinism and its naturalistic legacy, and a way of understanding divine action—ID thus intersects science and theology.

It is clear from Dembski's words that there is a conflation of science and metaphysics and an anti-evolution ideology. This has been the clear presupposition of those who promote ID on the public platform, such as Phillip Johnson, where evolution is equated with "naturalistic" evolution, a materialistic force rather than as a scientific mechanism to be embraced by science if it meets proper scientific criteria.

Michael Behe provides the scientific background to ID. He maintains that ID is a purely scientific approach, without ideological overtones. This aspect has been widely canvassed in public discussions. Behe makes much of the "irreducible complexity" of certain biological mechanisms. He asserts that certain biological processes are too complex or multifaceted to be explained by conventional science and therefore it is necessary to insert design as a mechanism. The examples given by Behe include the action of flagella ("bacterial tails"), the human eye and the complex chemical blood coagulation mechanism. He describes in great scientific detail the nature of some of these mechanisms maintaining that many consist of different components that could not have been derived by evolutionary mechanisms, but only by postulating ID to fill the gaps. ID is thus based on the *scientific* description of complex systems and the *unscientific* assertion that science cannot explain these further without postulating ID. We thus have a contradiction of terms. ID cannot be tested scientifically and yet it is presented as a scientific mechanism for life processes. There is therefore a clear confusion of science with metaphysics.

There is little encouragement therefore from the mouths of its founding fathers that ID is a purely scientific theory. This is why ID as science has been denounced by scientific bodies, not on ideological grounds but on scientific methodological grounds. It is simply a reiteration of the long discounted concept of a "god of the gaps." It is the argument that our scientific ignorance today leaves a place to be filled by God—a god of the gaps.

Such a god becomes smaller and smaller with each advance of science. Gaps in science however are to be filled by doing more and better science, not by postulating a god to fill them.

Many of the examples of "irreducible complexity" described by Behe have been exploded by subsequent scientific work, as has been discussed by Kenneth Miller (who is both a prominent Catholic layman and eminent US biological scientist) in his book *Finding Darwin's God*.

In summary then, I would maintain that the ID position is flawed philosophically, scientifically and theologically:

- *Philosophically*, it confuses metaphysics (religious belief) with physics (science). The Christian (and indeed Jewish and Muslim) doctrine of creation asserts that God is the creator and sustainer of the cosmos. This worldview contrasts with that of secular humanists who deny divine action. Scientists who hold either of these worldviews however can (and do) maintain that the exploration of nature in all of its complexity should be explored by scientific methods.
- *Scientifically*, it is flawed in that it proposes a "secondary cause" that cannot be tested scientifically.
- *Theologically*, the god of ID is a "god of the gaps," a cog in the machine, rather than the Christian God of the Bible or of the creeds—"God the creator of heaven and earth." Christians are thus presented with a theology that sells God short—a God who gets smaller with each advance of science.

Darwin, Biology and Genetics

63. Adapting to the impact of Darwin

By Stephen Ames

Darwinism is seen by many as the alternative to any religious view of the world, but atheism and creationism are not the only possibilities.

Over 150 years ago Charles Darwin introduced a revolution in our view of life on planet Earth, with his book, *On The Origin of Species,* ranked with the changes in worldview wrought by Copernicus and Galileo. It showed how the appearance of design in plants and animals could be produced by blind causal processes operating over a vast stretch of time without reference to any purpose or any miraculous creation by God. The revolution has impacted everyone, not least Christians and their faith.

Darwin was born at Shrewsbury in 1809. His father was a free-thinking physician, his mother a Unitarian from the wealthy Wedgwood family. Darwin engaged in his "hobby" of natural history while studying medicine at Edinburgh University, but he was unable to cope with performing surgery. His father then sent him to Cambridge University in 1828, to prepare for ordination in the Church of England, where he happily absorbed William Paley's books setting out evidence for Christianity and God's direct creation of species specially designed for their environment. These were required texts for all students.

In 1831 Darwin was offered an appointment as companion to the captain of the HMS *Beagle,* on its mission to chart the coast of South America. In five years the *Beagle* circumnavigated the earth. Darwin soon became the ship's naturalist, collecting specimens of all kinds of animals, most especially from the Galapagos archipelago in the mid-Pacific. This was a life-changing voyage for Darwin. His discoveries showed a world that did not fit with Paley's views, nor with the creationism of Milton's *Paradise Lost,* which he took with him on the *Beagle.*

Darwin returned to England in 1836, where the extraordinary findings from his journey opened the door to the highest echelons of British science. He set about making sense of the vast amount of data he had collected, finding time to marry his cousin Emma Wedgwood and live as a gentleman with their ten children in Downe House, Kent. After a lot of hard thinking, correspondence with many people, and sheer brilliance, he put together his theory of "descent with modification." It has three key ideas: descent from a common ancestor, random variation of features and natural selection. The theory was able to explain a huge amount of different kinds of data. It was this that convinced many scientists and other people of the fact of evolution. Darwin believed it could explain everything from the smallest organism to the higher mental faculties of human beings.

Darwin's proposed mechanism, the blind, mechanical process of natural selection acting on random variations, was criticised for scientific and philosophical reasons. Lord Kelvin, the leading physicist of the day, calculated that the age of the earth was too short for Darwin's theory to work. Many strong supporters of evolution still sought some sense of purpose in nature and a natural base for moral values, which they could not find in blind natural selection.

Crucial to the eventual triumph of Darwin's ideas was the mid-nineteenth-century work of Gregor Mendel, a Moravian monk, on the basic principles of heredity—Mendelian genetics—from his study of peas grown in the monastery garden. His work was unknown to Darwin. Mendel's work was "discovered" in the early twentieth century and was eventually shown to fill in the gaps in Darwin's theory. By the 1930s this was put on a sound theoretical basis and one of the people responsible for this work was the Cambridge statistician and devout Anglican, R. A. Fisher, who later came to work for the CSIRO in Australia and is buried at St Peter's Cathedral in Adelaide. "Neo-Darwinism" is the modern synthesis of molecular biology and Darwin's theory, and this in turn is integrated into science's evolutionary cosmology, an account of the universe from the "big bang" to a "big freeze" future—the current best scientific view.

Darwin renounced Christianity, partly because of the New Testament teaching about the eternal punishment of unbelievers, which he found morally offensive. The break came after the death of his father in 1848 and of his ten-year-old daughter, Annie, in 1851. While he declined into agnosticism in his old age, I don't think Darwin was an agnostic when he wrote the *Origin*. He could still speak of the greater grandeur of the idea of God working through laws of nature (but not miraculously intervening in nature) to produce the extraordinary structures of diverse forms of life we see.

Darwinism and Christianity

So what about Darwinism and Christianity? There have been many anti-Darwin Christians since the nineteenth century. There is too great a

Adapting to the impact of Darwin

difference between their literal reading of the Bible and Darwin's view of the world, now developed to evolutionary cosmology from the big bang to today. Many people, such as Richard Dawkins, take Darwinism as the alternative to any religious view of the world.

But atheism and creationism are not the only possibilities. From the second half of the nineteenth century to today there have been many people, leading theologians and eminent scientists, who held together Darwinism and Christianity, with varying degrees of adjustment of beliefs, from a "liberal" to a recognisably "orthodox" faith represented by the creeds and the Scriptures.

Here are some of the issues Darwinism raises for students with regard to religion and Christianity in particular. First, life evolves according to blind causal processes and therefore without purpose. Nature is a "blind watchmaker," as Dawkins said. Second, the evolution of life on planet earth by natural selection is a costly process with a vast amount of pain and death—the problem of natural evil. On both points the kind of world we live in is felt to be very different from what you would expect if the world was purposefully created by a truly good God. This leads to a third difficulty: why would God create and keep the universe in existence and then use evolution by natural selection to bring life into existence? And why do so in a universe that, so it seems, will die a cold, dark death?

Another perceived difficulty is that religion is fixed, whereas science is dynamic, initiating new experiments and observations, testing theories and developing new theories as needed. Empirical inquiry is dynamic, whereas religion holds a set of static beliefs that will become outdated in the face of new factual discoveries by science. The "outdated" Bible is a standard example. This leads to the view that faith is "blind faith" held in the face of contradictory evidence. For some students, because Darwinism did away with Adam and Eve, there was no fall, no original sin and so no need of Christ. Others find Christianity is too anthropocentric and that human beings created in the "image of God" are not part of nature. This goes completely against the grain of Darwinism.

The "outdated" Bible

The easiest issue to deal with here is that of the Bible being "outdated" by science. My response is to recall Galileo, who showed how his telescopic observations and his Copernican view could be reconciled with the Bible. Drawing on the theological tradition, especially St Augustine, he proposed the "two books" principle. The "two books" are the "book of nature," and the Bible. For Galileo, God is the author of both books. It follows that ultimately there cannot be any contradiction between the two books. If one finds a contradiction, one or both books are being misread.

If we accept that God is the "author" of two books, the book of nature and the book of Scripture, and that the natural sciences do show us something of the book of nature, then it follows that the two books cannot be in contradiction when each is interpreted correctly, because there is only one author. If we come to an assured view about some part of the natural universe through our scientific work, how should we interpret parts of the Bible that seem to present a different account of the natural world? We will need to seek a different interpretation. But this should not mean simply interpreting the text in terms of the latest science: for example, saying "the six days are long epochs and the order of the plants and animals in Genesis is the same as for evolution." This avoids thinking about the meaning of the text and it certainly empties the text of its magnificent historical and theological meaning: for example, calling Jews to live faithfully contra the ideology of Babylon; the whole world being created effortlessly and peaceably by God speaking, not violently as in Babylonian stories about the gods; human beings created in the image of God, not as play things of the gods. This is not the only creation passage in the Bible, so we should be informed by the full witness of the Bible as valuable in forming our theological vision of the world from creation to consummation, and thus for our conversation with those deterred from Christianity by the sciences.

The problem of scientism

But there is a deeper issue. It is that a scientific view of the world is often confused with naturalism, or "scientism." This view not only accepts what the sciences say there is, but goes on to say "and that is all there is." That claim goes well beyond the methods and results of the natural sciences into a philosophy of nature. For many people, Darwinism has always been a materialistic view of life, a philosophy of nature and as such is inevitably in conflict with Christianity and other religions. Darwin thought of his theory as materialistic, in the sense of appealing only to natural causes, to explain the evolution of life including the higher mental capacities of human beings. But he wanted to exclude questions about ultimate beginnings and endings, which would take him into philosophy and theology.

Adapting to the impact of Darwin

Intelligent design (ID) is the most well-known current attempt to challenge this naturalism. I think it fails because in its present form it trades on "gaps" in what science has so far explained. Notoriously such gaps close with the advance of science and some of ID's gaps have been closed. Is there an alternative? Yes.

First, elaborate a Christian vision of the divine economy or plan for the whole creation revealed in Christ (Eph 1:10; 3:9–11). Ensure this is theologically rich enough to address the issues noted above, and show the theological significance of Darwinism in this larger setting. This can and must also include a principled place for open-ended scientific inquiry.

Secondly, identify those aspects of our human life that logically cannot fit within a naturalistic view of the world. Here I think of what it is to be a "person," the reality of genuine, non-reciprocal altruism of "good Samaritans," and the way human beings conduct inquiry, including that of the natural sciences. Hard thinking is needed, but something harder is to live a full human life in a materialistic culture without imbibing this naturalism. Can Christian communities offer that possibility?

64. Anglican encouragement led to Darwin's science

By David Young

24 November 2009 marked the 150th anniversary of Charles Darwin's famous book *On the Origin of Species*. Melbourne University zoologist David Young reflects on the circumstances that led up to this landmark publication as well as its Anglican context.

Looking at Darwin's *On the Origin of Species* in the context of his day gives us a valuable insight into his achievement and what it means for theology.

A good starting point is provided by two quotations that Darwin placed opposite the title page of *On the Origin of Species*. One comes from Francis Bacon's *Advancement of Learning* published in 1605. In this quotation, Bacon uses the traditional metaphor of God's two books to argue that we should study the book of God's works (nature) as much as the book of God's word (Scripture). Such an approach clearly encouraged the growth of science in a Christian civilisation.

The success of the physical sciences is reflected in the second quotation Darwin used. This comes from William Whewell's book on *Astronomy and General Physics* published in the 1830s. In the sentence quoted, Whewell says that as far as the physical world is concerned, we can see that God works through general laws and not through isolated cases of divine intervention.

The study of the living world was also encouraged and really took off during the seventeenth century. An important first step was that many species of plants and animals were collected and carefully described. For example, John Ray published his *Ornithology* in 1676, which described all the birds currently known across Britain and Europe.

In addition, Ray could see that each species of bird was well adapted to a particular way of life. He went on to use such adaptations as evidence for design by a creator in his book on *The Wisdom of God* published in 1691. The argument from design became a key part of the rational defence of religion in the eighteenth century and it encouraged the study of living things. However, it excluded the idea that God might work through natural laws, unlike the situation in physics and chemistry.

This perspective was widely accepted in England when Darwin was growing up in the 1820s. As a young person, he developed a passion for natural history and was described by his uncle as "a man of enlarged

Anglican encouragement led to Darwin's science

curiosity." These qualities led to his studying at Cambridge University with a view to becoming a clergyman in the Church of England. That made sense because the positive relation between the church and science often resulted in the clergy pursuing science as a major hobby. John Ray was an early example.

In fact, some clergy were academics who contributed to the growth of science at English universities. William Whewell was a clergyman on the staff at Cambridge when Darwin went there to study. So too was Adam Sedgwick (a geologist) and John Henslow (a botanist).

These clergymen encouraged Darwin's interest in science, especially Henslow, with whom Darwin developed a close friendship. This circumstance, he later recalled, "influenced my whole career more than any other." It was through Henslow that Darwin received the invitation to accompany Captain Fitzroy on HMS *Beagle*.

Now a significant transition was taking place in science at the time Darwin was on his *Beagle* voyage. Thanks to the new science of geology, it was becoming clear that great changes had occurred in the history of the earth and that these must have involved vast periods of time. These changes included the disappearance of some remarkable animals that had previously inhabited the earth.

The first fossil finds of ichthyosaurs, plesiosaurs and dinosaurs were published during the 1820s. Finds like these showed that the world had once been dominated by creatures that were now utterly extinct. Successive groups of animals had evidently been driven to extinction and then been replaced by others.

Geologists like Adam Sedgwick thought that the earlier groups had been rendered extinct by sudden upheavals in the earth's geology. New groups of animals had then been designed and introduced by divine acts of "creative interference." Such views were challenged by the publication of Charles Lyell's three-volume *Principles of Geology* in the 1830s.

There was no need to invoke sudden upheavals in Lyell's view. Present-day causes acting gradually over millions of years could explain most geological changes, including extinctions. But if extinctions happened gradually over long periods of time, then where did the new species come from to replace the old ones? Lyell acknowledged this question but had no suggestions to offer for an answer.

This was how the matter stood when Darwin arrived home from the *Beagle* voyage in 1836. He soon took up the species question, as it was called, and early in 1837 became convinced that new species arise by evolution. He began a series of notebooks on the topic and developed the idea of natural selection as a cause of evolution in 1838. He then began work on the details and penned two private essays in 1842 and 1844.

In the public arena, Darwin was forestalled by Robert Chambers, who published a popular book, *Vestiges of the Natural History of Creation*, in 1844. This advocated a divinely ordered process of evolution as the obvious explanation for new species. The *Vestiges* caused a sensation but was easily dismissed by specialists like Sedgwick and Whewell. Darwin saw the need to do much better and went on with his species work.

Another person who became convinced of evolution was Alfred Wallace, who published a paper on "the introduction of new species" in 1855. His paper summed up the distribution of species in space and time and clearly hinted at a natural origin for new species. Then in 1858 he sent Darwin a draft of another paper, which not only proposed evolution but also suggested natural selection as a likely cause of evolution.

Friends arranged for a joint publication of this paper along with some of Darwin's material. Darwin was then galvanised into writing *On the Origin of Species*, which proved to be a superb culmination of twenty years' work. It was essentially "one long argument" skilfully aimed at contemporary naturalists and so made evolution convincing in a way that it had never been before.

It is important to note that Darwin was successful because his book effectively answered the species question. It was not a matter of ideology or politics. In the introduction to the sixth edition, Darwin could say that now "almost every naturalist admits the great principle of evolution." Of course concerns were raised about its implications for theology but these were not on the same scale as the rumpus that followed publication of the *Vestiges*.

Some Anglicans in Darwin's time were concerned that God had been removed now that design was not involved. Others saw that God was perfectly capable of working through natural laws. Among the latter were clergy ranging from Charles Kingsley to Frederick Temple, who eventually became Archbishop of Canterbury. Thus the Anglican Church both encouraged the science that led up to *On the Origin of Species* and was soon able to accommodate its conclusions within the fold.

65. Divine activity through evolution long accepted

By David Young

How does God fit in with evolution? Fortunately, a satisfactory answer was worked out long ago and evolution has never been a big problem within the Anglican Church.

One might get the impression that evolution was a big problem at first, thanks to the famous clash between Bishop Samuel Wilberforce and Thomas Henry Huxley. This encounter took place at a meeting of the British Association for the Advancement of Science held in Oxford in July 1860. Darwin's theory of evolution came up for discussion and Wilberforce spoke against it, while Huxley sprang to its defence.

However, this colourful exchange was not the big match it is often made out to be. Wilberforce was not speaking in an official capacity on behalf of the Church of England but gave his opinion as an individual who often took part in meetings of the British Association. He raised some genuine scientific difficulties for Darwin and was much concerned about the implications of human evolution.

The Huxley/Wilberforce debate did excite some public comment in the media but this dried up after 1860. Then in the 1880s and 1890s, the topic was resurrected by people with a secularist agenda. It was these later interpretations of the Oxford debate that gave rise to the myth of it having been a big clash between science and the church.

At the time, Wilberforce's speech was part of a range of responses to Darwin within the Anglican Church. As it happens, a very different point of view was put forward on that same weekend in Oxford by the Rev. Frederick Temple. He was chosen by the university to preach to the British Association in a sermon on the Sunday and accordingly made some comments on the relationship between science and religion.

Although he did not mention Darwin by name, Temple was seen as making room for the idea of evolution in a Christian context. He firmly put aside the habit of linking theology to things that science could not explain. Instead, he emphasised that God could be seen to act through the laws of nature and was not an alternative to them.

Viewing the laws of nature as an expression of divine activity was not a new idea thought up in response to these fresh developments in science. It was a well-established idea, which had its roots in the doctrine of creation worked out by the early church. Early Christian theologians had come to the conclusion that God created the world out of nothing and not out of any pre-existing matter.

This concept clearly implies that the world depends entirely on God for its existence and cannot exist all in its own. From this perspective, the world's continuing existence needs divine action just as much as its original creation in the beginning. In fact, theologians thought that this ongoing maintenance of the world could be regarded as a continuing creation.

So, with the growth of modern science in the seventeenth century, the newly discovered laws of nature were readily seen as maintained in being by divine activity. Such a view was expounded by Robert Boyle among others. Initially, this idea of a divinely maintained world did not include the possibility of change from the original condition.

It was only with the advent of geology at the end of the eighteenth century that evidence of changes in the history of the Earth began to mount up. Among the most remarkable changes was the extinction of many species and their replacement by new ones over time. The question of where these new species came from was raised in the 1830s, especially by Charles Lyell's *Principles of Geology*.

An early response to this question was provided by the astronomer, Sir John Hershel, who wrote to Lyell in 1836. He acknowledged that the origin of new species was currently a mystery but thought it should turn out to be a natural process. This was because all previous scientific studies led us to think that the creator worked through laws of nature.

Thus, Temple's sermon in 1860 matched the views of some contemporary scientists as well as having deep roots in Christian theology. However, there were one or two obstacles to be overcome before such a view could be generally accepted. The main one was the argument from design as set out in William Paley's influential book, *Natural Theology*, published in 1802.

The aim of natural theology was to provide evidence from nature for the existence of God, without reference to Scripture or theology. Paley did this by drawing a parallel between the adaptations of living organisms to their way of life and human artefacts such as clocks. Since the latter are clearly the products of design rather than chance, he argued, this must surely be true of the former as well.

Species of living organisms were thus interpreted as machinery that had been designed and manufactured by God. Unfortunately, this

Divine activity through evolution long accepted

eighteenth-century imagery placed divine creativity in direct competition with any natural laws that might generate new species. In turn, this meant that God was thought to intervene in his own creation to create a succession of new species during Earth's long history.

When Darwin published his *Origin of Species*, it was obvious that evolution was in conflict with this interpretation of God as designer and manufacturer of species. This has made it easy for those with an axe to grind to claim that evolution has removed God from the world and to link evolution with atheism. But Darwin himself did not adopt such a position; as he wrote to a friend, "my views are not at all necessarily atheistical."

An important point here is that this image of God as designer and manufacturer was also inconsistent with a full creation theology. As we have seen, theology had long recognised natural laws as an expression of divine activity and not a competing agency. That is why people were soon able to accept that species had not been created by divine intervention but rather created by God working through evolution.

This changed perspective on God's creativity was well under way by the time Temple returned to Oxford to deliver the Bampton Lectures in 1884. Entitled *The Relations Between Religion and Science*, his lectures welcomed Darwin's contribution to science and to theology. By showing that natural laws prevailed in biology, just as they did in physics and chemistry, Darwin had restored the unity to be expected in God's creation.

While other concerns might remain, reflections such as these effectively dealt with the central question of how God could fit in with evolution. As to the standing of such views within the Anglican Church, we can best refer back to that earlier weekend in Oxford. Of the two clergy who spoke on issues related to evolution, it was Frederick Temple and not Samuel Wilberforce who became Archbishop of Canterbury.

66. Genetics: One way God speaks to us

By Chris Mulherin

To what extent do our genes determine who we are and the choices we make, including our sexual preferences? Phil Batterham, a professor of genetics at The University of Melbourne, explored this question in a public lecture as part of science week in St Paul's Cathedral, Melbourne. "Genetics: The language of God?" was the theme of the week which was held in August 2014 to coincide with National Science Week.

Professor Phil Batterham is President of the International Genetics Federation and a Christian believer who thinks that his science and his faith "fit together pretty neatly."

In a fascinating lecture on the basics of genetics along with some theological commentary, Batterham led us through the genetic revolution since Francis Crick and James Watson's discovery of the structure of DNA in 1953.

Is genetics the language of God, as the original lecture title asked? It's not "the" but "a" language of God, said Batterham. God speaks in so many ways and in many languages; genetics is just one of them.

Batterham started and finished his talk relating science and faith. He affirmed his confidence in the key attributes of God: that God exists, that God is eternal, omnipresent and the creator. And although God's method might be debated, the creation of life includes the genetic code, which lies at its core. And God is also the author of truth, all truth.

As for science, its key attributes are the pursuit of truth about the earth, life and our universe. So both science and faith are on the same page; they both seek honestly to pursue the truth to the best of their ability.

The implications of this joint pursuit of truth are that for the believing scientist, the truths of science, said Batterham, should be "incorporated into our theology and they give us an expanded view of God's purpose in this world."

DNA *and genetics: the basis of life*

Genes are the building blocks of life; they are a code made of molecules strung together in spiral ladders of DNA (deoxyribonucleic acid), and these

DNA ladders are found coiled up in the cells of living organisms. For humans, those DNA ladders have about three billion rungs and add up to a total length of two metres per cell, with coiled copies found in chromosomes at the heart of almost every cell in our bodies.

Now, human beings have about ten billion billion cells (of some 200 different types) and, if you add up that two metres of DNA found in most cells, your total DNA uncoiled would stretch some twenty billion kilometres.

This coded information found in the human genome is inherited from our biological parents and from the single cell that we all started out as. And it is this DNA template that directs the multiplication and differentiation of cells, ensuring that *you* turn into a *you* and not a me or a fish or giraffe.

Did my parents make me do it?

The genetic code also controls development of our bodies from conception to adulthood. It determines gender and eye colour; it regulates functions of our body such as digestion. And it also plays a part in our behaviour, abilities, lifespan, height, sexuality ... the list goes on.

If all the raw material that went into making the original "me" came through genetic inheritance, and genes contribute so much to who we are, that raises the vexed question of nature and nurture. How much do environmental, cultural and historical factors affect the "outcome" that is me? And how much freedom do I have to choose to be different to the way I've been made and moulded?

Batterham happily waded into the nature and nurture question and in one sense his answer was clear. While genes determine the raw material and might affect dispositions to be or behave in one way or another, this is conditioned by various factors such as nutrition, education, family traditions and values.

Nurture

For some characteristics, these environmental factors can have a huge influence and sometimes an entirely environmental predisposition might seem inherited. Batterham suggested we think of the keen Collingwood Footall Club supporter; if you draw up the family tree and trace the Magpie "trait" through the generations, you might well conclude (wrongly) that there is a gene for following the 'Pies.

So, says Batterham, while lots of traits have a genetic base—he mentioned alcoholism, schizophrenia, Alzheimer's disease, autism, major affective disorder, reading disability—that doesn't mean that they aren't also environmentally influenced. In fact, most traits that we care about are a result of nature *and* nurture, including for example, both obesity and high blood pressure. And sifting nature from nurture is not easy. That's where identical twins are a great help because both twins have identical genes. So any differences between the two must be due to environmental factors and not due to their inherited (identical) DNA.

Thorny theological questions

Batterham also dealt carefully but clearly with questions that are causing a stir in theological circles today: that of sexual preference, that of our first human progenitors and a comment on biologically caused suffering.

On human sexual preferences: Batterham suggested that there is strong evidence for a genetic component. In organisms from flies to mammals genetic variation influences sexual preference, and twin studies reinforce this view.

But he warned against thinking that the balance between nature and nurture was the same in every person. The question of why someone has a gay orientation, for example, has a person-specific answer. For some it could be totally environmentally influenced, for others it could have a high genetic component. And in either case, he said, it doesn't mean that the person's choices do not also play a part.

On the human family: Part of the power of the genetics revolution is that it not only offers an understanding into genetically inherited traits, but it is also a window into the past history of the species. Because we inherit our genes from our parents (and they from theirs, and so on ...) genes give a historical record of human lineage on this planet.

We now know that the origins of modern humans lie somewhere near the Rift Valley in Kenya. At that time, only a few thousand generations past, all humans were indigenous Africans. From there, humans started moving out about 60,000 years ago, and Australian Indigenous people were one of the first groups to arrive in their current location.

So, Batterham explained, the idea of "human races" is "a significant social construct with no biological basis." Any two humans on earth are 99.9 percent identical in DNA sequence. That one part in one thousand that differentiates us into unique individuals is totally insufficient to justify biological separation into races; humans are, genetically speaking, one race.

"We are one large extended family," he said, and "while we are all very similar, apart from identical twins, we are all unique. God made us to be one family; God made us to be unique. God embraces and loves diversity, so we should too."

On suffering: Batterham closed his talk with a restrained doxology on the problem of suffering; how can the loving creator God allow suffering to be encoded in some people's DNA? This conundrum is older than the discovery of the spiral helix by Watson and Crick, said Batterham, "but I make sense of this apparent conflict by reflecting on the Christian hope that our life is eternal and our life on the planet is not a millisecond in comparison with that eternity."

67. Simon Conway Morris: Human race not the result of "dumb luck"

By Murray Hogg

Simon Conway Morris, Professor of Evolutionary Palaeobiology at the University of Cambridge, and an Anglican Christian, believes that human beings are not a cosmic fluke and that we are almost certainly the inevitable outcome of the evolutionary process.

Here's a question to ponder before you read any further: What does science tell us about the origin of human beings? Was your answer: "The scientific evidence shows that our evolutionary emergence is virtually guaranteed"? It's not surprising that you had to read that twice, or three times, for it is the surprising answer given by Simon Conway Morris in lectures he gave in Australia in 2009.

Now, Conway Morris is no quack of questionable scientific credentials. In addition to being Professor of Evolutionary Palaeobiology at the University of Cambridge and a leading expert on early life on earth who has undertaken research in countries around the world, he is also a Fellow of the Royal Society, he has been awarded the Walcott Medal of the National Academy of Sciences and the Lyell Medal of the Geological Society of London. He has given the Royal Institution Christmas Lectures, the highly prestigious Gifford Lectures (University of Edinburgh) and was the opening speaker at the Biological Evolution Facts and Theories Conference at the Pontifical Gregorian University, Rome. He has appeared alongside thinkers such as Alister McGrath and John Polkinghorne in "Test of Faith"—a Christianity and science study resource from the UK's Faraday Institute. And he previously visited Australia to study its local fossils and to take up the Australian Academy of Science's Selby Visiting Fellowship awarded to "distinguished overseas scientists" who are "outstanding lecturers to the general lay public." Finally, he is, I have to say, a genuinely nice guy to boot.

In his lectures he emphasised two central ideas. The first is this: of the vast array of outcomes apparently possible in evolution, only a small few can actually work. We can be certain, he quips, that given the universe of evolutionary possibilities we still know that pigs can't fly! The second is that this limited range of possibilities means that evolution repeatedly "converges" on similar solutions to evolutionary problems. This is the idea

Simon Conway Morris: Human race not the result of "dumb luck"

of evolutionary convergence and is now an accepted part of evolutionary theory. The problem is, as Conway Morris points out, that most evolutionary theorists expect quite the opposite! As Stephen Jay Gould once memorably remarked: if we could re-run the tape of life we would arrive at remarkably different outcomes. Yet Conway Morris marshals an impressive array of evidence to show precisely the opposite. Against all expectations, the evolutionary lotto draw results in the same numbers time and time again, suggesting that evolution is not so much a gamble after all.

What is particularly interesting is that Conway Morris' approach is thoroughly scientific. He starts with the scientific data and risks no sweeping metaphysical claims about their implications. Here we may contrast him with the "creation science" movement, which is quite open about the fact that they start with a particular interpretation of Genesis which guides their scientific deliberations. Others, such as ultra-atheist Richard Dawkins, go beyond science when they claim that the data can prove, or disprove, the existence of God. Conway Morris, by contrast, offers a third way which shows far more respect for the scientific method, one which accumulates a broad body of evidence, and restricts himself to inferring from it the best scientific explanation.

Before going further, it will be helpful to note that Conway Morris' book on this subject (*Life's Solution*, Cambridge University Press) runs to over 450 pages, including more than 100 pages of scholarly references and footnotes. His case depends on a vast range of examples covering the entire spectrum of the life sciences (as he said to me, most people's problem when it comes to understanding evolution is they don't appreciate just how vast is the subject). So I hope the reader understands that this brief chapter is the throwing of a tiny pebble against which Conway Morris' full effort is an overwhelming, full-scale artillery barrage of scientific fact after fact after fact. His lectures could hardly do the full subject justice, this chapter certainly doesn't, and readers wishing to make a full appraisal should consult his book.

Perhaps Conway Morris' most interesting example is that of the camera eye—the sort of eye we humans have in which light passes through an "aperture" (the iris) and is focused onto a "film" (the retina) by the lens. This enormously complex structure, with all its associated biochemical and neural processing systems, must certainly be an evolutionary once-off, right? Well, no. It turns out that evolution has used this solution more than once and it is found in such unrelated species as ourselves, octopuses, snails, marine worms, and even some jelly-fish! The only possible conclusion is that of convergence: that evolution has stumbled across this highly complex solution more than once. Shear dumb luck, it seems, is far from an adequate explanation.

Other intriguing examples abound: tool use amongst humans, chimps, dolphins and certain species of birds; self-awareness in humans, chimps, elephants and birds; the remarkable convergence of "universal music" in whale and bird song. Other examples include social ordering, communication, intelligence, cooperation in hunting, a range of biochemical capabilities such as the ability to withstand sub-zero temperatures or to draw energy from sunlight via photosynthesis, and a myriad more. That such things have evolved multiple independent times is beyond question. What is in question, and it's a question Conway Morris raises again and again, is how one explains the fact that the same restricted group of phenomena evolve again and again in quite unrelated lineages? On the theory that the outcomes of evolution are a matter of dumb luck—and again we are back to Stephen Jay Gould's famous metaphor of the re-running of the tape of life—we simply wouldn't expect such a narrow set of frequently repeated phenomena.

So what conclusions does Conway Morris draw from the above? Well, he is far too good a scientist to confuse the scientific and theological enterprises and of his scientific findings the most he will say is that they are "congruent" with his Christian faith. But, as an Anglican Christian whose favourite author is G. K. Chesterton, Conway Morris is not afraid to count himself amongst the orthodox. He affirms the incarnation, death and resurrection of Christ. And when it comes to the bigger questions of life, of the significance of humans and our place in the universe, his attitude seems to me summed up in the closing words of his 2005 Boyle Lecture at St Mary-le-Bow Church in Cheapside, London:

> Science, when it treats creation as a true creation, and thereby faces up to its responsibilities, may well be important ... It seems ultimately, however, that it is the knowledge and experience of the incarnation, the wisdom and warnings given by Jesus in the Gospels, and not least the resurrection that in the final analysis are all that matters.

Afterword

Faith in a World of Science

By Rodney Holder

Dr Rodney Holder is Emeritus Course Director of The Faraday Institute for Science and Religion, Cambridge, UK, and is a Fellow Commoner of St Edmund's College, Cambridge. He read mathematics at Trinity College, Cambridge, and was awarded a DPhil in astrophysics and later a theology degree from Oxford University. He is a priest in the Church of England. His books include *Big Bang, Big God: A Universe Designed for Life?* (Lion Hudson, 2013).

Introduction

As emerges very powerfully from this wide-ranging collection of essays and interviews, we have to express our Christian beliefs today in a credible way to a world in which science is a significant part of the cultural backdrop. This raises significant questions for faith. How do we understand God's revelation in Scripture in the light of science? How is our understanding of God, of creation and the human person affected by the findings of science? Can we still recite the creeds with integrity? Or is our faith undermined by modern developments, such as in the cognitive study of religion? Is there still a place for natural theology or has that been undermined by Darwin? Can Christian faith claim to be rational in the way that science is understood to be rational and based on empirical evidence? What resources do we have in Scripture and the Christian tradition to address these issues? And, on the other hand, what can faith contribute to science? As at the time of the Reformation, I shall argue, picking out and reflecting briefly on some of the topics that have gone before, that Christians can indeed return *ad fontes*, to the biblical sources and classical exponents of the faith, for insight. Inevitably, some important topics will pass without comment. Most notable, perhaps, is climate change and creation care, but then, this topic among many others is ably and extensively covered in the main text.

God talk

I have often been asked how my science has affected my view of God. As described in this book, people often have a quite erroneous view of how Christians think of God in the first place, and they think that somehow

A Reckless God?

science demolishes that picture of God. Perhaps their image of God is as an old man in the sky with a long white beard. It seems to me that there was great wisdom in the Hebrew prohibition on images of God. If that ever was our picture of God, then it certainly shouldn't have been.

I have answered that question in two different ways in different contexts. The first is to say that my view of God is the one of classical Christian theism, so in that sense my view of God hasn't changed. God is necessary being, eternal and self-subsistent, omniscient, omnipotent and perfectly good, and is the creator of all that is not God. Why would science change that view?

God as necessary being explains why the universe, which is contingent, exists. The universe may or may not have existed and could have been different from what it is. The idea of God as necessary being explained the existence of the universe for St Thomas Aquinas and it does that today, and scientists such as Lawrence Krauss, who say that God is redundant because the universe can create itself, are mistaken. Indeed, as explained in the main text, Krauss is talking obvious nonsense when he says that the universe can create itself out of nothing. He gets the attention he does for saying it simply because he is an eminent scientist—but of course that does not mean that he knows any philosophy or theology. No, God is needed as creator today just as much as when St Thomas formulated his five ways or "proofs" of God back in the thirteenth century.

Of course Scripture does not use sophisticated philosophical terms to describe God, but it seems to me that this concept of God is implicit in Scripture. Some of the great passages about God as creator can certainly be read that way. "Have you not known? Have you not heard? The Lord is the everlasting God, the creator of the ends of the earth" (Isa 40:28). God is "the high and lofty One, who inhabits eternity" (Isa 57:15). And of course there is much that can be drawn from God's naming of himself to Moses at the burning bush: "I am who I am." At the end of the Bible God says, "I am the Alpha and the Omega, who is and who was and who is to come, the Almighty" (Rev 1:8). What the great theologians such as St Thomas have done is taken this biblical understanding, revealed through the narrative of Scripture, and systematised it, expressing it in the most sophisticated philosophical language available to them.

On the other hand, there is some discussion today in science–religion circles, and this emerges in the book, as to whether we need to qualify the notions of God's omnipotence and God's omniscience. Even classical

Faith in a World of Science

theologians would say that God cannot do what is logically contradictory, or is in contradiction to the character of God, and similarly God cannot know what is not there to be known. An important debate today concerns whether God knows the unformed future, which is dependent on the actions of free creatures, and, related to that, whether God self-limits his power by making such creatures. I shall not delve into these controversial points too deeply, but merely say that in any case God knows all possibilities, and, analogously to the perfect chess grandmaster in Peter Geach's analogy, who wins the game whatever the opponent plays, God will bring about his purposes and cannot be thwarted.

When it comes to the problem of theodicy, a couple of the authors of this book commend rejection of God's impassibility. Indeed, it seems that the biblical God does have feelings and there is much to commend Jürgen Moltmann's view that the Father suffers the loss of the Son on the cross, marking the most terrible and incomprehensible rent in the perfect union of the persons of the Trinity. Jesus' cry of dereliction—"My God, my God, why have you forsaken me?"—is drawn to our attention a couple of times in the main text, most movingly by Moltmann himself as he describes his own journey to faith. I return to the problem of theodicy below.

My second answer to the question of whether my view of God has changed is in any case much less complicated and philosophical, and it echoes much of what is said in this book. It is to say, yes, my view of God has been enlarged and enhanced by science. That is often not the answer people expect. They think science has displaced God, a view ably refuted in this book.

The Psalmist looked up at the heavens and said, "When I look at your heavens, the work of your fingers, the moon and the stars that you have established, what are human beings that you are mindful of them?" Does knowing what a star is diminish my view of God? Our sun is a very ordinary star, but it's a gigantic nuclear reactor, 860,000 miles across and generating 400 million million million million watts of power. There are more than 100 billion stars in a typical galaxy and more than 100 billion galaxies in the observable universe. Does knowing this diminish my view of God? Of course not. It tells me that God is far more majestic than I can imagine.

So why might people think that our view of God has been diminished by science? One reason is that they think the laws of nature somehow supplant God: We don't need God to explain things in nature any more because the laws of physics and the processes of organic evolution do that for us. People

A Reckless God?

think that God cannot act in the world because the laws of nature determine all that happens—whether the laws are deterministic or not. But, again as explained in this book, that is to make a separation between nature and the God of nature. God is the one who endowed his creation with the laws which science discovers and God acts through the laws of nature.

Again we have tremendous resources in the Christian tradition to deal with this point. St Augustine said: "God has established in the temporal order fixed laws governing the production of kinds of beings and qualities of beings and bringing them forth from a hidden state into full view, but his will is supreme over all. By his power he has given numbers to his creation, but he has not bound his power by these numbers."[1]

The "numbers" here refer to what we would call laws of nature, and God works through the laws with which he has endowed the creation but is sovereign over those laws, whether or not he limits himself by allowing creaturely freedom.

St Thomas Aquinas is similar to Augustine, but, as described in the main text, he makes a distinction between primary and secondary causes. God is the primary cause of all things—he it is who causes them to exist and gives them the powers they have to act. But creatures can and do act as secondary causes through the powers with which they have thus been endowed by God (*Summa Theologiae*, 1a, 105.5). Just like Augustine, Aquinas sees the development of a human person from the womb to adulthood in terms of God's action, but God acting in and through the processes with which he has endowed nature. Interestingly enough, Darwin in his *Origin of Species* makes a similar comparison between the creator bringing about the production and extinction of species through secondary causes and the birth and death of an individual through secondary causes.

For Aquinas, as for Augustine, this does not limit God's action because God freely created the secondary causes and they are subject to him. Again, this is philosophical language but what is being described is implicit in Scripture. God is portrayed as creator and sustainer of the universe and the one who controls the elements. God is the one who gives the fruits in their season, and Jesus' divinity is revealed to his disciples when the wind and the waves obey him.

---

1 *The Literal Meaning of Genesis (De Genesi ad Litteram)*, Ancient Christian Writers, vols 1 and 2, translated and annotated by John Hammond Taylor SJ (New York and Mahwah, NJ: Paulist Press, 1982), Book VI, 13, 23, p. 194.

Faith in a World of Science

It seems to me that we have great resources in the tradition with which to counter the arguments of our atheist opponents or answer the questions of genuine enquirers.

Holy Scripture

As we have seen in this book, some Christians and some atheists, are locked into a narrow literalist reading of Scripture, and this can be a great stumbling block to the way we talk about our faith in the context of science. That applies particularly, but not exclusively, to the way we read the early chapters of Genesis. But yet again we have great resources in the tradition, and it is important to see how the great theologians of the early church interpreted Scripture.

Origen, writing about 200 AD, demonstrates the absurdity of a literal interpretation in many places and he gains warrant for an allegorical interpretation from New Testament writers' interpretation of the Old Testament. Regarding the early chapters of Genesis, he writes: "Now what man of intelligence will believe that the first and the second and the third day, and the evening and the morning existed without the sun and moon and stars?"[2]

Augustine made the same point another 200 years later. He added the further point that evening and morning only succeed each other at a particular location; the earth as a whole experiences both simultaneously. There is a particularly helpful passage in Augustine, which is worth quoting in full as an object lesson to the modern day biblical literalist:

> Usually even a non-Christian knows something about the earth, the heavens, and the other elements of this world, about the motion and orbit of the stars and even their size and relative positions, about the predictable eclipses of the sun and moon, the cycles of the years and the seasons, about the kinds of animals, shrubs, stones, and so forth, and this knowledge he holds to as being certain from reason and experience. Now, it is a disgraceful and dangerous thing for an infidel to hear a Christian, presumably giving the meaning of Holy Scripture, talking nonsense on these topics; and we should take all means to prevent such an embarrassing situation, in which people show up vast ignorance in a Christian and laugh it to scorn. The shame is not so much that an ignorant individual is derided, but that people outside the household of the faith think our sacred writers held such opinions,

---

2 *On First Principles (De Principiis)*, Gloucester, MA: Peter Smith, 1973, IV, III, 1, p. 288.

and, to the great loss of those for whose salvation we toil, the writers of our Scripture are criticized and rejected as unlearned men. If they find a Christian mistaken in a field which they themselves know well and hear him maintaining his foolish opinions about our books, how are they going to believe those books in matters concerning the resurrection of the dead, the hope of eternal life, and the kingdom of heaven, when they think their pages are full of falsehoods on facts which they themselves have learnt from experience and the light of reason?[3]

Amen to that! Augustine is saying that we can and should accept the findings of reason and experience—what we now call science—and it is seriously damaging to the gospel to interpret the Scriptures so as to contradict these findings. Of course at the time of the scientific revolution thinkers such as Francis Bacon and Robert Boyle, who saw God as writing the two books spoken of in the main text, that of nature and that of Scripture, realised that these books could not contradict each other because they came from the same author.

Faith and Reason

The new scientific atheists set up a false dichotomy between faith and reason. Science is rational, based on reason and empirical evidence. Religion is based on faith, which, according to Richard Dawkins' definition "means blind trust, in the absence of evidence, even in the teeth of evidence."[4]

Alister McGrath has challenged Dawkins on his own ground and asked the question, "Okay, then, where is the evidence for Dawkins' definition of faith?" He finds there isn't any and this is simply Dawkins' own definition of faith—he's setting up a straw man to attack.

Philosopher and theologian Keith Ward quotes a classic Dawkinsian caricature of religion: "one of the truly bad effects of religion is that it teaches us that it is a virtue to be satisfied with not understanding ... If you don't understand how something works, never mind: just give up and say God did it." Ward's response is scathing: "I have to say that this is one of the most obviously false statements in the history of human thought."[5] In

---

3   *De Genesis ad Litteram*, Book I, 19, pp. 42–43.
4   Richard Dawkins, *The Selfish Gene*, (Oxford: Oxford University Press,1989 [1976]), 198.
5   Keith Ward, *Why There Almost Certainly is a God: Doubting Dawkins*, (Oxford: Lion Hudson, 2008), 61.

reality of course it is the opposite way round: belief in God has provided the motivation to *find* understanding. Again, this highlights Dawkins' ignorance of history. And again we can go back to classical exponents of the faith such as Anselm who said, "I believe in order that I may understand" and who talked of "faith seeking understanding."

The modern-day theologian Wolfhart Pannenberg's understanding of faith is interesting in comparison with Dawkins: "a person does not come to faith blindly, but by means of an event that can be appropriated as something that can be considered reliable. True faith is not a state of blissful gullibility."[6] For Pannenberg, the main point is that God has acted in history through the life, death and resurrection of Jesus Christ and we have publicly accessible evidence in the form of witness testimony to those events, which can therefore be judged reliable.

I agree with Pannenberg about this. But my own way of thinking about faith is the same as a way that appears in this book, namely in terms of a relationship of trust, like marriage, rather than in terms of belief without evidence. I have faith in my wife. When we married I had a certain amount of evidence and that has grown over the years, though it would be a funny kind of marriage in which I deliberately kept doing experiments to find out if she was still trustworthy. Similarly there is evidence for Christian belief, and that is important when it comes to sharing our faith—for having a reason for the hope that is in us, as Peter says. But we don't go putting God to the test all the time. That would be a problem, for example, with attempting to set up "controlled" experiments to test whether God answers prayer.

Notwithstanding the warning about putting God to the test, there is in fact plenty of good evidence on which to base one's faith and this is where I see the importance of natural theology.

Natural theology

Aquinas' five ways can be considered a form of natural theology. Indeed natural theology embraces the traditional arguments for the existence of God, especially the cosmological and teleological arguments.

---

6 Wolfhart Pannenberg, ed., with Rolf Rendtorff, Trotz Rendtorff, and Ulrich Wilckens, *Revelation as History*, trans. D. Granskou, (New York: Macmillan, 1968), 138. (1st German edition, *Offenbarung als Geschichte*, Göttingen: Vandenhoeck and Ruprecht, 1961).

A Reckless God?

From Aquinas to William Paley in the nineteenth century there was a subtle shift in natural theology. Aquinas gave general arguments whereas the scientific revolution brought in arguments based on the particular. In his famous book, *Natural Theology, or Evidence of the Existence and Attributes of the Deity, Collected from the Appearances of Nature* (1802), Paley gave the famous example of a watch found on a heath. The watch, which possessed great intricacy, was obviously designed. How much more so the eye observing it? This form of the argument was undermined by Darwin, as noted in this book, but not, I would argue, the form Aquinas deployed.

For Aquinas and others, natural theology has been a preliminary for revealed theology. It seems to me that natural theology is useful both for removing barriers to belief in God and providing positive reasons for belief; it provides the groundwork for the more specific and important belief in Jesus Christ as Lord and Saviour.

In recent years, philosophers of religion, most notably Richard Swinburne in Oxford, have built up a cumulative case for the existence of God based on the cosmological and teleological arguments, the existence of consciousness and morality, the evidence from history and miracles, and the evidence of religious experience. For Swinburne each piece of evidence makes it more probable that God exists and the clinching piece of evidence making God more probably existing than not is that from religious experience. For me, as for Swinburne, John Polkinghorne and others, the fine-tuning of the universe, discussed in this book, is an important component of such a cumulative argument. In contrast to scientific explanation in terms of scientific laws and initial conditions, theistic explanation is explanation in terms of personal agency.

I just said that this provides the preliminary to revealed theology. However, I believe that the division between natural and revealed theology is a somewhat artificial one, especially because I think we need to justify accepting what is purported to be revelation, "to give a reason for the hope that is within us," in Peter's words. Indeed, that task is more important today than ever. Maybe suspicion of the claims of revelation began with the rise of biblical criticism in the nineteenth century, but that suspicion is very much alive today: witness the attacks on the Bible from Dawkins and his friends.

From a Christian point of view the most important fact to justify in Scripture is the resurrection of Jesus. And I agree with Pannenberg that we have enough to go on in terms of the evidence for the resurrection to accept it as true. The resurrection does not conflict with science because it is a unique event. We believe, not for scientific reasons, but for the kinds of reasons we believe much else—the testimony of witnesses.

There has been a great deal of work in recent years on the resurrection, with excellent books produced by Tom Wright,[7] Richard Swinburne[8] and others making this case. There's a very good article on it by Lydia and Timothy McGrew in *The Blackwell Companion to Natural Theology*.[9] Evidence for the resurrection has traditionally been a big part of Christian apologetics and I think should remain so. It may not cut much ice in the postmodern context, but in the context of science, which believes in objective truth and is about truth-seeking, it adds a parallel search for objective truth in the historical and theological realm.

The problem of theodicy

This book makes a helpful contribution to that most perplexing of all the problems that face us, namely that of theodicy. I certainly cannot solve the problem myself in such a short space, and it may not be soluble this side of the eschaton in a long space either. However, it is worth pointing out that as Christians, and as Christians who are scientists, again we have resources, and there is something we can say.

The book of Job, drawn to our attention several times in the chapters above, is one of the most profound reflections on human suffering in all literature. Job is the archetypal innocent sufferer. His suffering is graphically described. He is so disfigured that he is abhorred by all around him, family and friends alike.

I was reminded of Job when reading Afghan author Khaled Hosseini's immensely moving book *And the Mountains Echoed*. In the book Thalia is a young Greek girl whose face has been savagely mutilated by a dog. Thalia faces the taunts of schoolchildren, and her own mother insists she wear a mask—not for Thalia's sake, but to save her mother's embarrassment. Thalia ends up being adopted. Her friend and adoptive brother, Markos, eventually becomes a plastic surgeon. Interestingly, Thalia later refuses plastic surgery when Markos offers it, because her face has by now become her identity:

---

7 N. T. Wright, *The Resurrection of the Son of God* (London: SPCK, 2003).
8 Richard Swinburne, *The Resurrection of God Incarnate* (Oxford: Oxford University Press, 2003).
9 Timothy McGrew and Lydia McGrew, "The Argument from Miracles: A Cumulative Case for the Resurrection of Jesus of Nazareth," in William Lane Craig and J. P. Moreland (eds.), *The Blackwell Companion to Natural Theology* (Chichester: Wiley-Blackwell, 2012 [first edition, 2009]), 593–662.

A Reckless God?

"It is what I am," she says. Thalia is a quietly heroic figure in the novel who comes to terms with her disfigurement and manages to lead a fulfilling life caring for her adoptive mother, even if some avenues are closed off to her. Thalia is a modern day Job, an innocent sufferer, of which there are of course all too many in our broken world.

Job's so-called "comforters" present reasons to him for why he is suffering. They say he has sinned and is reaping the reward from God of his own evil doing. This diagnosis fails in the case of Job and of course it fails in all the cases of innocent suffering down the ages. In the New Testament it is repudiated by Jesus when he is called upon to heal a man blind from birth (John 9).

Against these taunts, Job protests his innocence. He longs to be brought to the heavenly court, to meet God and to state his case. What is remarkable is Job's conviction that he will be vindicated in the end. Indeed, the very familiar verses Job 19:25-27, often spoken at funerals, stand out in the Old Testament as an affirmation, in the midst of the most intense suffering, that he will be vindicated: "For I know that my redeemer lives, and that at the last he will stand upon the earth; and after my skin has been thus destroyed, then in my flesh I shall see God, whom I shall see on my side, and not another."

Job is expecting a redeemer, an advocate, to appear for him and to present his case in the heavenly court. A redeemer in ancient Israel is one who set a slave free from bondage, or one who regained family property to keep it within the family, or one who married a widow to preserve an inheritance for the woman's late husband, as happens in the book of Ruth.

The language Job uses, of a redeemer to represent him in the heavenly court, brings to mind not just ancient Jewish legal customs but what the New Testament has to say, especially in the epistle to the Hebrews. Hebrews 7:25 tells us that our Lord Jesus Christ is "able for all time to save those who draw near to God through him, since he always lives to make intercession for them." And Hebrews 8:1 tells us that "we have a high priest, one who is seated at the right hand of the throne of the Majesty in the heavens, a minister in the sanctuary." Jesus is our redeemer and advocate in the heavenly court.

Hebrews also tells us that Christ himself is like Job, the archetypal innocent sufferer. It was fitting, we are told, that the pioneer of our salvation should be made perfect through suffering (2:10). Moreover, he had to become like his brothers and sisters in every respect, so that he might be a merciful and faithful high priest in the service of God, to make a sacrifice of atonement for the sins of the people. Because he himself was tested by what he suffered, he is able to help those who are being tested. (2:17–18)

Another archetypal figure in the Old Testament is the Suffering Servant who appears four times in the middle section of the book of Isaiah:

He had no form or majesty that we should look at him, nothing in his appearance that we should desire him. He was despised and rejected by others; a man of suffering and acquainted with infirmity; and as one from whom others hide their faces he was despised, and we held him of no account. Surely he has borne our infirmities and carried our diseases; yet we accounted him stricken, struck down by God, and afflicted. (Isa 53:2–4)

One of my favourite paintings is the Isenheim altarpiece in Colmar, Alsace. Painted by Matthias Grünewald between 1512 and 1516, it depicts, unusually for its time, a harrowing, twisted and bloody picture of Christ on the cross, his body covered in sores. The painting was displayed by monks to patients suffering from St Anthony's fire, a terrible disease which wracked the body in the way Christ's body is wracked in the painting. It would have been seen every day when these sufferers came into the chapel for services. It is truly a picture of Christ entering into our own human sufferings and thereby bringing redemption. It depicts Christ directly fulfilling the prophecy of Isaiah 53:4: "Surely he has borne our infirmities and carried our diseases; yet we accounted him stricken, struck down by God, and afflicted."

There is no simple or easy answer to the problem of pain and suffering. But what our Christian faith offers is a God who comes alongside us, who in the person of his Son bears the pain of a human life lived for others, but is rejected by those he came to save. Jesus sits at the right hand of the Father as our advocate and the Holy Spirit transforms us from the inside so that we are prepared for the eternal life beyond the grave which Christ's death and resurrection secure for us.

The catholic creeds

I want to end with one of the questions I raised in my introduction, that is, whether today we can still recite the creeds with integrity as those of us who are Anglicans, at least, do every Sunday. The creeds are considered to mark out orthodox Christian belief from heresies, for example committing us to belief in God as Trinity, Father, Son and Holy Spirit. My answer is a resounding yes: we can recite the creeds with integrity as John Polkinghorne explains brilliantly in his book *Science and Christian Belief*.[10]

For example, there is very good reason indeed to believe in God as creator, as affirmed in the first clause of the Nicene Creed: "We believe in one God, the Father, the Almighty, maker of heaven and earth, of all that is, seen and unseen."

When it comes to the Son, we affirm a lot about who he is and we also say what he has done: "For our sake he was crucified under Pontius Pilate; he suffered death and was buried. On the third day he rose again in accordance with the Scriptures." I have already spoken about the absolutely vital place of the resurrection, and it is the resurrection which makes sense of the rest of what is said about Jesus.

The creed ends by saying "We look for the resurrection of the dead, and the life of the world to come." Again, the resurrection of Jesus is what guarantees our own resurrection. Of course we couldn't know these things unless God had revealed himself to us, and indeed the idea of a crucified God is a stumbling block to Jews and folly to Gentiles. But in fact Christ did die and rise again, and we have good, solid evidence for it.

So has modern science affected the way we talk about faith? Yes, and no. Yes, we do express that faith in a new context. Learning from the great theologians of the past, we interpret the Scriptures in a way that is compatible with the discoveries of science. Natural theology still has an important role to play, but natural theology takes a different form from that which it took in John Ray and William Paley (both mentioned in the text). And, if anything, our view of God should be enhanced rather than diminished by the discoveries of science. Science paints an awesome picture of the universe and God is certainly free to act in the world he has created. Indeed, when all is said and done, we can still hold fast to the abiding truths of the Christian faith "uniquely revealed in the Holy Scriptures and set forth in the catholic creeds, which faith the Church is called upon to proclaim afresh in each generation."

---

10 John Polkinghorne, *Science and Christian Belief: Theological Reflections of a Bottom-up Thinker* (London: SPCK, 1994).